Cognitive Systems Monographs

Volume 41

The Cognitive Systems Monographs (COSMOS) publish new developments and advances in the fields of cognitive systems research, rapidly and informally but with a high quality. The intent is to bridge cognitive brain science and biology with engineering disciplines. It covers all the technical contents, applications, and multidisciplinary aspects of cognitive systems, such as Bionics, System Analysis, System Modelling, System Design, Human Motion Understanding, Human Activity Understanding, Learning of Behaviour, Man-Machine Interaction, Smart and Cognitive Environments, Human and Computer Vision, Neuroinformatics, Humanoids, Biologically motivated systems and artefacts, Autonomous Systems, Linguistics, Sports Engineering, Computational Intelligence, Biosignal Processing, or Cognitive Materials—as well as the methodologies behind them. Within the scope of the series are monographs, lecture notes, selected contributions from specialized conferences and workshops, as well as selected Ph.D. theses.

Indexed by SCOPUS, DBLP, zbMATH, SCImago.

More information about this series at http://www.springer.com/series/8354

Diego Azevedo Leite

The Twenty-First Century Mechanistic Theory of Human Cognition

A Critical Analysis

Springer

Diego Azevedo Leite
Federal University of Alfenas
Alfenas, Minas Gerais, Brazil

ISSN 1867-4925 ISSN 1867-4933 (electronic)
Cognitive Systems Monographs
ISBN 978-3-030-63679-1 ISBN 978-3-030-63680-7 (eBook)
https://doi.org/10.1007/978-3-030-63680-7

This Springer imprint is published by the registered company Springer Nature Switzerland AG
The registered company address is: Gewerbestrasse 11, 6330 Cham, Switzerland

Acknowledgments

I became interested in debates about major theoretical accounts of human cognition since I got to know some of them for the first time during the first semesters of my bachelor's degree in the field of psychology. Since this time, I could not stop thinking about it and directing my research towards it, in a way or another. I find such topics extremely challenging and complex (after all, they reflect the complexity of human cognition and human behavior), but, at the same time, they are very interesting for me. This book is, therefore, also the result of many years of reflection about particular theoretical and empirical issues concerning central dimensions of contemporary theories of human cognition present in cognitive science.

A great amount of effort and dedication was necessary for concluding this work. It is no simple task, indeed, to analyze and compare systematically major theories in the field of cognitive science. The task, however, is a very significant one, as I see it. Thus, I tried to start this project, bringing different theoretical ideas together and evaluating their internal and external consistency and coherence without losing sight of my final objective: a critical evaluation of one particular proposal. During the development of this work, I tried to find unity, maintaining the clarity, objectivity, accuracy and rigor, as well. Hopefully, the final result will be useful for the reader.

Many people made very important contributions to the project of publishing this book, which is an updated and revised version of my doctoral dissertation, finished in 2018. I would like to thank, firstly, Prof. Sara Dellantonio at the University of Trento (Italy), Prof. Saulo de Freitas Araujo at the Federal University of Juiz de Fora (Brazil), and Prof. Achim Stephan at the University of Osnabrück (Germany), for their help, advice and suggestions concerning the development of my doctoral dissertation. They have shown me how to do academic work with rigor and high-quality, and this was important in many ways for the construction of my dissertation—they certainly have had a very significant impact on the formation of my own professional thinking.

Equally important is to express my gratitude to the *Department of Psychology and Cognitive Science* at the University of Trento, where I conducted my doctoral research with a scholarship for the entire period. I thank also the *Institute of Cognitive Science* at the University of Osnabrück, where I spent my period abroad for the doctoral program, which was a great experience for me.

I would like to thank Steffen Achenbach, Rondineli Mariano, Agnese Marcolla, Nan Yang, Jodi Guazzini, Gregor Hörzer and Helena Frenzel, as well, for the very interesting conversations we had about my research topics, and also for their support. All of them contributed, directly or indirectly, to this project. A special thanks goes to Prof. Eric Batista Ferreira at the Federal University of Alfenas (Brazil) and to Prof. Osvaldo Frota Pessoa Junior at the University of São Paulo (Brazil) for their support and advice concerning the publication of this book.

I would also like to extend my gratitude to my Springer Publishing Editor, Mayra Castro, for her help, attention, assistance and collaboration. I am grateful too for all the support and assistance from the people working in the Springer editorial team—they helped me in many ways so that the book could be published.

Finally, I would like to express my deepest gratitude to my family, for their continuous support in all dimensions of my life.

Diego Azevedo Leite

Contents

Abbreviations

BDTHC	Belief-Desire Theory of Human Cognition
CTHC	Computational Theory of Human Cognition
DNTSE	Deductive-Nomological Theory of Scientific Explanation
DSTHC	Dynamical Systems Theory of Human Cognition
MCTHC	Molecular and Cellular Theory of Human Cognition
MTCBS	Mechanistic Theory of Complex Biological Systems
MTGC	Mechanistic Theory of General Cognition
MTHC	Mechanistic Theory of Human Cognition
MTSEBS	Mechanistic Theory of Scientific Explanations in Biological Science
MTSECS	Mechanistic Theory of Scientific Explanations in Cognitive Science
TSHC	Theory of Situated Human Cognition
UTHC	Unified Theory of Human Cognition
UTSECS	Unified Theory of Scientific Explanations in Cognitive Science

Chapter 1
Introduction

Human cognition (or mind)[1] has been considered throughout history until present days as one of the most intriguing and fascinating aspects of our reality, as well as one of the most interesting and important objects of scientific and philosophical inquiry. However, it is also a very complex and difficult object of study, not just because of its intrinsic particularities, but also because of its complicated and puzzling relationship with the human brain.

Human cognition and its relationship with the human brain are topics being investigated for centuries, resisting many efforts of a complete understanding (cf. Crane and Patterson 2000; Harnish 2002; Leahey 1997). Nevertheless, many developments within the natural sciences and specific areas of philosophy took place in the twentieth century and the beginning of the twenty-first century (cf. Bermúdez 2014; Chipman 2017; Kim 2011; Ochsner and Kosslyn 2014; Reisberg 2013; Samuels et al. 2012; Stephan and Walter 2013; Weiskopf and Adams 2015). These developments generated a great amount of expectations, enthusiasm and hope in the scientific and philosophical community towards the possibility that a better knowledge of brain processes would provide the understanding of the 'link' between the human brain and cognition—finally eliminating all the mystery surrounding it. Strictly associated to these developments, one of the main current strategies to solve this traditional problem of relating human brain and cognition is to assume the hypothesis that human physiochemical neural phenomena are somehow 'fully responsible for producing' human cognitive phenomena. According to this hypothesis, an objective and accurate scientific explanation of the neural mechanistic operations will show that neural processes somehow 'generate' all cognitive processes.

The pinnacle of these developments and explanatory strategy is arguably the articulation of a new theory of scientific explanation describing scientific practice

[1] I will use 'mind' and 'cognition' as synonymous in the present work. I will also use 'cognitive science' and 'psychology' as synonymous in the present work, except when I indicate otherwise. Finally, the terms cognitive 'phenomena' and 'phenomenon' will be used as cover terms for the natural regularities related to cognitive entities, events, processes, activities, operations, capacities, functions etc., cf. Weiskopf and Adams (2015, p. 2).

D. A. Leite, *The Twenty-First Century Mechanistic Theory of Human Cognition*, Cognitive Systems Monographs 41, https://doi.org/10.1007/978-3-030-63680-7_1

especially in the biological science (e.g. Bechtel and Abrahamsen 2005; Bechtel and Richardson 2010; Glennan 1996, 2002, 2017; Krickel 2018; Machamer et al. 2000; Woodward 2002) and the attempt to apply this theory of explanation to cognitive science (e.g. Bechtel and Wright 2009; Bechtel 2007, 2008, 2009a, b, c, d, 2010, 2012, 2016, 2017, 2019; Boone and Piccinini 2015; Craver 2007; Kaplan 2011, 2017; Milkowski 2013, 2016; Milkowski et al. 2018, 2019; Piccinini and Bahar 2013; Piccinini and Craver 2011; Piccinini 2007, 2012, 2015; Thagard 2006, 2009, 2018, 2019; Wright and Bechtel 2007; Zednik 2018). Thus, the theory is applied to cognitive capacities, including all human cognitive capacities.

In order to provide solutions for issues concerning scientific explanations of complex biological phenomena, this neo-mechanistic framework proposes what I will call a *Mechanistic Theory of Complex Biological Systems* (MTCBS), which includes a *Mechanistic Theory of Scientific Explanations in Biological Science* (MTSEBS); and in order to provide also solutions for the issues concerning human cognition, the new mechanistic philosophy offers also what I will call a *Mechanistic Theory of Human Cognition* (MTHC), which includes a *Mechanistic Theory of Scientific Explanations in Cognitive Science* (MTSECS).[2] Despite of dealing with different issues, MTHC and MTCBS are strictly related, since the latter provides foundations for the former. These foundations are provided in two senses: firstly, in an ontological sense, since MTCBS offers an account of biological mechanisms which can be used in order to understand natural cognitive systems; and, secondly, in an epistemological sense, since MTSEBS offers an account of scientific explanations suited to the biological sciences and considers cognitive science to be one of these. Therefore, in order to completely understand MTHC, a profound analysis of MTCBS is necessary.

However, my central concern in the present work is with MTHC. More importantly, the concern is solely with the particular formulation of the theory as applied to human cognition (i.e. human adult, normally functioning, cognition). The extension of a possible *Mechanistic Theory of General Cognition* (MTGC) is not entirely clear, i.e. whether the theory is so general as to be applied to all nonhuman animals capable of some kind of cognitive capacities and even to cognitive artificial systems, providing thereby a general and highly inclusive fundamental theory of all kinds of cognitive natural and artificial entities. Nevertheless, this issue of the extension of

[2]This taxonomy is mine and I am using it because I find it more rigorous than other loose taxonomies generally found in the relevant specialized literature, which can easily lead to confusion. Frequently, these mechanistic theories are loosely referred as 'mechanistic explanation', or 'mechanistic framework', or 'mechanistic philosophy', or simply 'mechanism'. In these cases, it is often obscure to what kind of issues exactly the framework is being applied, to what kind of theories, or scientific domains. Given this, I will use this new taxonomy because of two major reasons. Firstly, I want to express clearly the difference between different dimensions of application of the twenty-first century mechanistic philosophy of science: as applied to science, generally, and as applied to cognitive science, particularly; in this way I can express clearly what MTCBS and MTHC are concerned with. Secondly, I want to use this more rigorous taxonomy in order to classify more precisely different comprehensive theories concerned with the same problems—i.e. general theories of complex systems and their scientific explanations—and make systematic comparisons between them and their theoretical elements.

the theory beyond the limits of human cognition is out of the scope of the present work.

The aim of MTHC is to explain human cognitive phenomena through brain mechanisms and their physio-chemical 'information processing' operations. Allegedly, MTHC shows success in explanations related especially to the memory system (specifically, for example, memory consolidation) and the perceptual system (specifically, for example, visual perception). The framework includes not just a theory of human cognition, but also a theory of the human neuro-cognitive relation, i.e. it provides a possible solution for the problem concerning how we should explain the connection between human neural and cognitive phenomena, thereby relating neuroscience and cognitive science in an integrated scientific framework (cf. e.g. Bechtel 2008, Craver 2007; Piccinini and Craver 2011). This human neuro-cognitive integration advanced by the theory involves, roughly, considering cognitive processes as a kind of neural processes, understood in terms of information processes—such processes can be decomposed and localized in the brain as parts of a multilevel complex neuro-cognitive biological mechanism. Moreover, these explanations incorporate many theoretical, empirical and methodological scientific achievements made at the end of the twentieth century and at the beginning of the twenty-first century.

Given this, the mechanistic theory is often presented as one of the major approaches, or the major approach, for explaining human cognition in the twenty-first century (cf. Bechtel and Wright 2009, p. 125ff).

The new ideas promoted by prominent neo-mechanists are having major recognition and adherence. Mechanistic explanations of human cognitive phenomena have recently become increasingly dominant in the field of philosophy of cognitive science (cf. Bechtel 2009a, 2010; Samuels et al. 2012). They are also having an enormous influence in the fields of cognitive neuroscience (cf. Bechtel 2008, 2009d; Ochsner and Kosslyn 2014) and cognitive science (cf. Anderson 2015; Bermúdez 2014; Goldstein 2015; Smith and Kosslyn 2014; Thagard 2018). Many authors in these and other fields are indeed enthusiastic about the framework. Craver and Kaiser, for instance, claim that "thinking about mechanisms enriches and transforms old philosophical debates" (2013, p. 125). The authors suggest, moreover, that neo-mechanists made a "revolution" which "replaced the last vestiges of the once-received positivist gestalt with a new mechanistic vision"; such a vision is "expressed in the very language in which scientists talk about their work" and is "sensitive to problems faced within mechanistic research programs in areas as diverse as biology, cognitive science, ecology, and neuroscience" (Craver and Kaiser 2013, p. 126). Many neo-mechanists see mechanisms almost everywhere in nature and in science, from cognitive psychology to the bottom levels of biology, and sometimes even beyond. Some authors, for instance, argue that the mechanistic framework can provide a consistent way for "building a unified science of cognition" and for integrating cognitive science and neuroscience (cf. Piccinini and Craver 2011, p. 283, 285). Boone and Piccinini (2015) even consider such new ideas as the new revolution in the cognitive and in the neural sciences.

Given this excitement, Fazekas and Kertész claim we are "in an era of mechanisms" and that there are "enthusiastic people around us saying that mechanisms are everywhere" (2011, p. 366). Weiskopf goes so far as to the point to claim that we are "in the midst of a mania for mechanisms" (2011, p. 313).

However, despite all the developments in the sciences and in philosophy, which provide grounds for the neo-mechanistic theory, the investigations concerning human cognition and its relationship with the human brain still encounter many difficulties. More particularly, there are still a great variety of terminological and interpretative disputes concerning even the most central concepts of the relevant sciences. For instance, disputes about how to characterize 'mind', 'cognition', 'mental/cognitive processes', 'mental/cognitive representations', 'cognitive information processing', 'cognitive computation', 'consciousness', 'behavior', 'environment' and how to relate each of them with neural physio-chemical vocabulary and data (cf. Kim 2011; Weiskopf and Adams 2015). Moreover, cognitive phenomena are extremely diversified. There are many cognitive broad capacities, such as 'perception', 'attention', 'memory', 'imagination', 'language', 'reasoning', 'consciousness', 'sensation' 'emotion', 'motivation', etc. and each of these broad capacities have also sub-capacities (cf. Anderson 2015; Ochsner and Kosslyn 2014; Reisberg 2013; Ward 2015). It is equally difficult in many cases to relate this cognitive vocabulary with the neuroscientific one. This on the one hand.

On the other hand, it is still not entirely clear how many parts of the brain work, what parts or regions are most relevant to explain a particular mental phenomenon and what are the most relevant 'levels' upon which this explanation should be grounded, e.g. the level of molecules and cells, or the level of neural networks (cf. Ochsner and Kosslyn 2014; Ward 2015). Moreover, how many different kinds of neurons there are and how exactly all these varieties of cells connect to each other are largely still open issues. Consequently, it is not currently clear as well how exactly the neuro-cognitive relationship should be spelled out. In the literature there is a proliferation of terms with meanings that are frequently superficial, ambiguous or obscure. For instance: the brain 'produces', or 'gives rise', or 'generates', or 'implements', or 'enables' cognition; the brain 'is responsible for', or 'is the basis of', or 'is the substrate of' cognition; cognition, in turn, 'arises', 'emerges', or 'comes from' the brain; cognition is achieved 'in virtue of features' in the brain.

A related difficulty is that there is a great amount of controversy concerning central research strategies to inquire about the relationship between human cognitive functions and neural structures, i.e. there is still a great dispute concerning competing fundamental macro-theories (that belong to highly inclusive research programs, or paradigms) of human cognition currently available (cf. Bermúdez 2014; Clark 2014; Chemero and Silberstein 2008; Dale et al. 2009; Walter 2014). More particularly, for instance, the central role of neuroscience and neural activity for the understanding of human cognition is still questioned by more traditional and by more recent approaches. In order to explain human cognition, such approaches emphasize the role of an autonomous computational level of human cognition, relatively independent on its physical implementation; or the role of complex and dynamical environments; or even the role of external media (such as computers and other technologies), which can allegedly influence human cognition in a particular way. Their proponents claim, roughly, that human cognition is actually broader than the brain, being able to be implemented in a variety of different physical structures, such as in a human being or in a robot. These positions stand in strong contrast with neuroscientific bottom-up approaches to human cognition. As a result, it is a matter of major contemporary disputes which currently available macro-theory and related research program aiming to provide an explanation to human cognition is more plausible.

Another matter of major dispute in this context is how the scientific fields interested in the cognitive and brain research (e.g. cognitive psychology, cognitive science, cognitive neuroscience, portions of neuroscience) are related to each other. In the literature, the boundaries and relations between these sciences are far from clear (cf. Chipman 2017; Miller 2003; Ochsner and Kosslyn 2014; Reisberg 2013). As a consequence, it is also not clear how one should understand more precisely the explanatory strategies and goals of these research fields and the theories, research programs and general assumptions adopted by them in order to better compare and evaluate their success as scientific fields of research on human cognition.

In attempting to provide solutions for the issues mentioned above, many neomechanists working in the field of cognitive science strive to formulate a general comprehensive theoretical system that could provide a well-grounded, unified and systematically integrated theory of human cognition, establishing at the same time how exactly it is related with neural activity, and how this neuro-cognitive system can be explained (cf. Bechtel 2008, 2009a, 2010; Bechtel and Wright 2009; Piccinini and Craver 2011; Boone and Piccinini 2015; Milkowski 2016; Thagard 2006, 2009; Zednik 2018). The mechanistic framework attempts to provide unification for cognitive science. It attempts to explain diverse phenomena through common causes (in this case, common mechanisms) postulated by a macro (highly-integrative) theory. It is specifically in this sense of unification (and integration) that the mechanistic theory of human cognition can be considered as a *fundamental theory*, or a *macro-theory* present in the domains of cognitive science and cognitive neuroscience. These terms refer here to a theory that is highly inclusive, i.e. it attempts to account for a great variety of phenomena and to provide guidelines for research (similarly to the fundamental/highly-integrative theories in the domain of physics).

However, given the issues discussed above and other obstacles, in cognitive science the difficulties related to achieving a fundamental theory of human cognition are enormous. In fact, the very possibility of constructing such a unified theory of human cognition, of its relationship with the brain, and of its explanation has been questioned (cf. Allen 2017; Dale et al. 2009; Dale 2008; Green 2015; Sullivan 2016; Teo 2010), and there are even authors who might think that given the complexities involved, this goal is absurd, like pursuing something that belongs only to the realm of phantasy. Knowledge in cognitive science, cognitive neuroscience and neuroscience has expanded impressively and it continues to do so (cf. Chipman 2017; Ochsner and Kosslyn 2014; Reisberg 2013). The degree of specialization of knowledge in these sciences is astonishing. Thus, it has sometimes been contended that a single integrative research program cannot succeed. The issue becomes more dramatic especially given the lack of systematic integration even within micro-theories directed to more particular cognitive capacities, such as particular kinds of perception (e.g. there are many very different models of different kinds of perception—visual, auditory, olfactory, etc.). With consciousness the issue of integration of theories is even more problematic.

This problem, however, needs to be considered carefully, because it must be determined whether we are dealing with a unified and integrated cognitive system, or with a variety of different kinds of systems that perform particular 'cognitive' capacities, i.e. some sort of language system, perceptual system, memory system that cannot be related to each other in a systematic and integrated significant manner. If that is the case, it also affects the relationship between programs of research in cognitive science, i.e. how particular research on perception, memory or language are supposed to be related to each other. It affects, more generally, the very idea of integration in cognitive science and what kind of integration it is realistically possible to achieve and seek for. Accordingly, on the one hand, one could pursue the project of developing an account of human cognition that is strictly unified and would have to be applied in the same sense to all subfields of cognitive science. On the other hand, one could think that it is more reasonable to develop an account of human cognition that is more sensitive to the disciplinary diversity found in the field and, thus, pursue a project along this line, i.e. without worrying so much if common theoretical elements are not frequently present in the accounts of the various cognitive capacities.

This issue of disunity, or evident general lack of unification, in traditional psychology and contemporary cognitive science is important. There were some attempts to propose some sort of (broader and stricter) unification (e.g. Churchland 1986; Gardner 1985; Newell 1990; Von Eckardt 1993). However, the issue is particularly difficult given the characteristic diversity and fragmentary nature in the study of human cognition (cf. Green 2015). The situation is not the one in which we have different theories being applied to different aspects of a given phenomenon—what often happens in the physical sciences. Rather, in cognitive science there are very different macro-theories in tough competition for providing an explanation of the very same phenomena. And what is even worse, the strong disunity is found not just *between* macro-theories but also *within* macro-theories, i.e. between theories at a middle-level and theories at a micro-level that belong to a given macro-theory.

Consequently, as Cummins (2000, p. 139) pointed out, until one solid fundamental framework gain prominence, we will still remain suspicious that most or perhaps all of them are mostly (if not entirely) flawed: "the prevailing bewildering diversity of theories tends to undermine confidence in any". Nevertheless, it is perhaps also important not to press too much and too hard for unification as things stand currently in the field, so that we do not take the risk of "forcing more unity than the data warrant" (Cummins 2000, p. 139). This could as well eventually lead to a great distortion of the truly explanatory aims of cognitive science in terms of its *explananda*, and of the sort of explanatory theory that is genuine for the required explanation of the phenomenon under consideration, i.e. its *explanans*. In this respect, equally important to consider might be not to allow for excessive plurality, undermining unity excessively, so that one develops the suspicion of 'scientific promiscuity', i.e. no clear limits or strong internal integrations are necessary and almost everything in the science is permitted—anyone can define human cognition and specific cognitive processes at wish and choose according to one's particular taste the best methodology to investigate her/his own creation.

Finally, the traditional mind-body problem, which is the ultimate background for all the difficulties mentioned above, has been an enormous source of controversy for centuries both in philosophy and in the sciences. The difficulties are various. To begin with, it is still largely debated how even to formulate the central questions and central problems (cf. Crane and Patterson 2000). There are debates related to what fields of inquiry are more likely to solve the problem, and objections related to the very possibility of solving the problem. There are as well questions related to how an acceptable answer would look like. At any rate, this problem is still considered by many philosophers and scientists to be extremely important for a variety of issues that most human beings care so much about. Issues, for instance, concerning freedom, responsibility, happiness and dignity of human beings. Therefore, the mind-body problem and its related particular problems (more interestingly, e.g. its implications for any project of a science of human mind/cognition) are regarded as very worth of serious and meticulous study, even if they constitute at the same time an extremely complex and hard topic.

All the issues discussed above make the task of providing a unified theory of human cognition and of its relation with the human brain an extremely difficult endeavour. Thus, there are still many challenges that remain for such a general project of integration in cognitive science as the neo-mechanistic one.

Therefore, to the extent that MTHC attempts to provide a comprehensive unified theory concerning human cognition and how to explain it, it needs to provide solutions for the difficulties mentioned above. However, any attempt to provide answers needs extremely careful consideration and evaluation. Not just because the issues are very important or extremely complex, but also because frequently they involve individual worldviews, philosophical prejudices, highly central beliefs, and related emotions and attitudes. Hence, as regards such issues, all the assumptions need to be critically and rigorously examined, all the proposed solutions need to be fully scrutinized, even, and perhaps especially, the most central ones. Moreover, there is no adequate, systematic and comprehensive analysis and evaluation of this new mechanistic theory specifically applied to human cognition and to cognitive science so far. Such analysis is indeed an original project, as well as extremely necessary, given the importance of the topic at hand and the influence of the framework on the contemporary cognitive and brain sciences. Since the theory can be considered both groundbreaking and extremely controversial, a theoretical analysis of it might help us to understand which of its aspects actually bring us forward in the understanding of the neuro-cognitive (or mind-body, or mind-brain) relationship, and which questions remain unanswered—needing, therefore, further investigation.

The present work, therefore, consists in a theoretical critical appraisal—a critical examination and evaluation—of the most central aspects of the twenty-first century mechanistic theory of human cognition and its relationship to the human brain. The ultimate goal is to offer a systematic analysis that will show the most ground-breaking and attractive aspects of the theory, on the one hand, and what are the aspects that need further improvement and elaboration, on the other. An evaluation like this can help us to distinguish in the proposed point of view what is sound from what is false or confused. It is necessary to discover if all the most important claims are rightful and if there are truly no ungrounded pretensions. The conceptual and discursive clarity of the theories must be addressed and evaluated. After all, what we want for our scientific and philosophical theories is "that unfeigned respect that reason grants only to that which has been able to withstand its free and public examination" (Kant 1781/1998, p. 101).

Specifically, I intend to analyze whether the neo-mechanistic account is indeed the best one currently available concerning human cognitive processes, and whether neo-mechanistic explanations are indeed the best kind of scientific explanation for human cognitive processes investigated by cognitive science. In other words, the central goal of the work is to scrutinize whether the neo-mechanistic framework is successful in providing a coherent, consistent and plausible unifying account of human cognition and of scientific explanations in the field of cognitive science.

In order to evaluate the plausibility of the neo-mechanistic framework, it is important, first of all, to understand its clearest formulation as a theory of human cognition and, at the same time, a theory about how to explain it. Moreover, it is necessary to clarify in what consists the integrative neuro-cognitive relationship present in the framework and to what extent this theoretical construction is—or is not—the most plausible available. As a way of balancing the positive and negative aspects of this theory and make a reasonable evaluation of its aspirations, I will also consider examples of applications of the theory to concrete human cognitive processes, as well as the major criticism raised by competitor theories of human cognition present in contemporary cognitive science. Finally, I will consider the implications of these debates for the integrative ambitions of the neo-mechanistic framework in cognitive science.

I will start the investigation, in Chap. 2, by exposing and considering the central ideas of the *Mechanistic Theory of Complex Biological Systems* (MTCBS). I will show the theory's most important aspects and attractive features according to its proponents. Nonetheless, it will not be my aim at this point to discuss critically the central ideas endorsed by the proponents of the theory, but rather to present the most consistent picture of the theory's central tenets, since there are still many unclear aspects related to the proposal and many debates about them in the specialized literature, even among its own proponents. The general purpose of the chapter, therefore, is to get the main ideas and commitments of the framework as correct, accurate, and precise as possible. This is the only way of undertaking a reasonable and fair judgment of its merits, as well as of what needs to be rejected and of what is important to be retained. More particularly, I will introduce the theory and the context in which it aroused in the final period of the twentieth century together with a brief overview of the philosophical foundations of this new twenty-first century version of mechanistic explanations of biological systems. After this, I will consider some philosophical debates concerning the nature and scope of different kinds of scientific explanations and discuss the central ideas related to the kind of explanations upon which the mechanistic proposal is grounded. Contrarily to the logical empiricists that dominated philosophy of science predominantly in the first half of the twentieth century, the new twenty-first century mechanists forcefully argue that the classic *Deductive-Nomological Theory of Scientific Explanation* (DNTSE)—which emphasizes the use of laws and logical deductions in scientific activity—cannot be applied with equal success to the life sciences, as biology, neuroscience, and (as they see it) cognitive science. The central scientific explanatory activity in cognitive science as well as in biology, so they argue, is the search for mechanisms and for a theory that explains how they work in nature.

In Chap. 3, I will provide a systematic and analytical exposition of the most central theoretical aspects of the *Mechanistic Theory of Human Cognition* (MTHC). I will show that the theory is clearly committed to a form of physicalism, on the one hand, but, on the other, it rejects certain kinds of traditional epistemological reductionist approaches. The framework attempts to offer a pluralist and integrative mechanistic view concerning the relationship between human brain and cognition, a view that is applied to phenomena and to theories overall in cognitive science and cognitive neuroscience. This general pluralist integrative neuro-cognitive relation is the most important pillar grounding the theory's application to human cognition; thus, it calls for careful and systematic consideration. Another point that demands careful examination is the suggestion that cognitive neuroscience is the best suited contemporary scientific enterprise to achieve this integration and ultimately provide all the explanations about human cognitive processes. After understanding the kind of explanation that the theory advances and its central commitments on the relationship between neural processes and cognitive processes, I will finally investigate how the framework is applied in concrete to two paradigmatic cases of human cognitive phenomena. The first case is related to the perceptual system, and the second case, to the memory system. In this way, it is possible to evaluate the application of the theory to particular psychological phenomena.

In Chap. 4, I will compare the neo-mechanistic theory with one of its major contemporary competitors, as I call it, the *Molecular and Cellular Theory of Human Cognition* (MCTHC). My aim in this chapter is to evaluate to what extent the main arguments presented by the proponents of MCTHC against the neo-mechanistic theory, directed to particular aspects of it, represent great threats to the aspirations of the neo-mechanists. I start with a characterization of the theory and the context of its rising. MCTHC supports a 'ruthless (strong) neuro-cognitive reductionism', as a form of scientific integration for cognitive and neural science, based on current neuroscientific work present in the field of molecular and cellular neuroscience. This theory presents a clear challenge to the neo-mechanistic theory, which is committed to causal and explanatory pluralism and a weak autonomy of higher-level sciences. After characterizing the neuroscientific reductionist position more precisely, I discuss the neo-mechanists' answer to the challenge and their attempt to stand with pluralism, instead of reduction. A meticulous analysis of their replies shows, however, that the challenge of explanatory reduction cannot be overcome with the arguments the neo-mechanists provide, and the theory, therefore, needs to be understood ultimately as reductionist.

I will analyze and evaluate, in Chap. 5, the major criticism to the neo-mechanistic account presented by leading proponents of two other theories, as I will call them,

Dynamical Systems Theory of Human Cognition (DSTHC) and the *Theory of Situated Human Cognition* (TSHC). I start with a general characterization of the first proposal and present the central challenge that it poses to the mechanistic theory, namely the thesis of the impossibility of decomposition and localization and inquiry into components of a system in isolation. This impossibility is due to the high degree of complexity and causal interaction in dynamical complex systems, such as the human neuro-cognitive system. After this characterization, I discuss the replies offered by influential neo-mechanists. These authors attempt to include complexity in their framework, so that this high degree of causal interaction will not present a problem, but will rather be compatible with their general thesis that a system must be organized in a particular way and this organization can be of different kinds. Accordingly, the organization of a dynamical complex system is just a kind of organization that a whole neuro-cognitive biological mechanism can have. Moreover, they argue that even to investigate highly complex dynamical systems some degree of decomposition and localization can be useful, since it can offer a preliminary step towards a dynamical explanation or help to analyze the phenomena in smaller portions, avoiding some extreme unattainable holism. In this way, the neo-mechanistic theory can be made compatible and assimilate dynamical systems theory in the way it is currently framed in cognitive science. The price paid, though, is the necessity to extend the applicability of the neo-mechanistic account to systems that cannot be fully decomposed, thus making decomposability a matter of degree.

Concerning TSHC, I provide, first of all, a general overview of this relatively new theory. As is well known in the literature, it includes different theses that can be hardly integrated into a single, comprehensive theoretical framework—and it is often considered extremely controversial. However, at least two versions of this view can be distinguished: a weak and a strong one. The weak version of the situated theory emphasizes the need of considering the causal interactions between the body and environment with human cognition, in order to better understand how cognitive processes occur. The strong version of the theory, however, holds that bodily, environmental and external media can also in many cases be part of human cognitive processes, as actual components in the cognitive whole system. I will concentrate in the second version of this view (the only one that presents some clear originality) and discuss to what extent the central arguments presented by the proponents of this theory generate difficulties for the neo-mechanistic view. After this, I discuss the replies of the neo-mechanists to these arguments. The conclusion I argue for is that this issue remains an open matter, since the neo-mechanists have not been able yet to overcome the arguments offered by the proponents of the situated theory. The arguments used by neo-mechanists in their most detailed reply do not touch the point at issue and, therefore, do not provide an answer for the challenges raised by the situated approach. Consequently, TSHC in its strong form remains an alternative to MTHC.

In the final chapter, I analyze and evaluate the criticism presented by proponents of two other main contemporary macro-theoretical approaches to human cognition in cognitive science, as I call them, the *Computational Theory of Human Cognition* (CTHC) and the *Belief-Desire Theory of Human Cognition* (BDTHC). I start with a brief presentation of these theories and in which respect they present challenges to MTHC. These two theories share many theoretical elements, but they also diverge on some important aspects. In the analysis, I show the similarities and differences between them to the extent that they are relevant for the discussion. I also use some genuine examples of concrete human cognitive higher-level complex phenomena in order to help in the clarification of some issues and the arguments presented. The central discussions here will concern two issues. Firstly, scientific explanations in cognitive science: particularly whether cognitive science explanations are provided by laws or by mechanisms; and whether they are always necessarily a kind of scientific mechanistic explanation. Secondly, the nature of human cognitive phenomena: i.e. it will be investigated whether the positions defended by influential advocates of MTHC on crucial issues—such as multiple realizability, type neuro-cognitive identity theory, cognitive representations, and cognitive computation—can provide indeed an integrative account concerning fundamental issues in order to unify cognitive science (or at least the majority of the field), and then unify cognitive science with the relevant portions of neuroscience interested in investigating human cognition. The analysis shows that the most influential advocates of MTHC are substantially divided concerning these major issues in cognitive science. Moreover, they present positions that turn out to be highly controversial and, in some cases, even misleading.

In my final remarks, I conclude that the proponents of MTHC have made important contributions for the advancement of many theories and discussions in cognitive science, cognitive neuroscience and philosophy of cognitive science. They offer a systematic analysis of the concept of mechanism and of what it means to provide mechanistic explanations in fields such as cognitive science and cognitive neuroscience. However, there are substantial differences in the very formulation of the theory by different influential authors and substantial disagreement between them on central issues. Moreover, many of the new positions offered by influential neo-mechanists on central issues are extremely controversial, misleading or need further elaboration. Therefore, MTHC, as it stands, cannot offer the ambitious revolutionary integration of cognitive science (a field particularly characterized by its diversity and fragmentation) that some of its most influential advocates promise. Nevertheless, I do not think that a substantial integration of the field (respecting its diversity) is impossible to be achieved, and I offer some suggestions for future work in order to reach this goal.

References

Allen, C.: On (not) defining cognition. Synthese (2017). https://doi.org/10.1007/s11229-017-1454-4

Anderson, J.R.: Cognitive Psychology and Its Implications, 8th edn. Worth Publishers, New York (2015)

Bechtel, W. Reducing psychology while maintaining its autonomy via mechanistic explanations. In: Schouten, M., Looren de Jong, H. (eds.) The Matter of the Mind: Philosophical Essays on Psychology, Neuroscience, and Reduction, pp. 172–198. Blackwell Publishing, Malden, MA (2007)

Bechtel, W.: Mental Mechanisms: Philosophical Perspectives on Cognitive Neurosciences. Routledge, New York (2008)
Bechtel, W.: Constructing a philosophy of science of cognitive science. Topics in Cognitive Science **1**(3), 548–569 (2009a)
Bechtel, W.: Looking down, around and up: Mechanistic explanations in psychology. Philos. Psychol. **22**(5), 543–564 (2009b)
Bechtel, W.: Mechanism, modularity, and situated cognition. In: Robbins, Philip, Aydede, Murat (eds.) The Cambridge Handbook of Situated Cognition, pp. 155–170. Cambridge University Press, Cambridge (2009c)
Bechtel, W.: Molecules, systems, and behavior: Another view of memory consolidation. In: Bickle, J. (ed.) The Oxford Handbook of Philosophy and Neuroscience, pp. 13–40. Oxford University Press, New York (2009d)
Bechtel, W.: How can philosophy be a true cognitive science discipline? Top. Cogn. Sci. **2**, 357–366 (2010)
Bechtel, W.: Identity, reduction, and conserved mechanisms: Perspectives from circadian rhythm research. In: Gozzano, S., Hill, C.S. (eds.) New Perspectives on Type Identity: The Mental and the Physical, pp. 43–65. Cambridge University Press, Cambridge (2012)
Bechtel, W.: Investigating neural representations: The tale of place cells. Synthese **193**, 1287–1321 (2016)
Bechtel, W.: Explicating top-down causation using networks and dynamics. Philos. Sci. **84**, 253–274 (2017)
Bechtel, W.: Resituating cognitive mechanisms within heterarchical networks controlling physiology and behavior. Theor. Psychol. **29**(5), 620–639 (2019)
Bechtel, W., Abrahamsen, A.: Explanation: A mechanist alternative. Stud. Hist. Philos. Biol. Biomed. Sci. **36**, 421–441 (2005)
Bechtel, W., Richardson, R.: Discovering Complexity: Decomposition and Localization as Strategies in Scientific Research, 2nd edn. MIT Press, Cambridge, MA (2010)
Bechtel, W., Wright, C.D.: What is psychological explanation? In: Symons, John, Calvo, Paco (eds.) The Routledge Companion to Philosophy of Psychology, pp. 113–130. Routledge/Taylor & Francis Group, New York (2009)
Bermúdez, J.L.: Cognitive Science: An Introduction to the Science of the Mind, 2nd edn. Cambridge University Press, Cambridge (2014)
Boone, W., Piccinini, G.: Cognitive neuroscience revolution. Synthese (2015). https://doi.org/10.1007/s11229-015-0783-4
Chemero, A., Silberstein, M.: After the philosophy of mind: replacing scholasticism with science. Philos. Sci. **75**(1), 1–27 (2008)
Chipman, S. (ed.): The Oxford Handbook of Cognitive Science. Oxford University Press, Oxford (2017)
Churchland, P.S.: Neurophilosophy. MIT Press, Cambridge, MA (1986)
Clark, A.: Mindware: An Introduction to the Philosophy of Cognitive Science, 2nd edn. Oxford University Press, Oxford (2014)
Crane, T., Patterson, S. (eds.): History of the Mind-Body Problem. Routledge, London, UK (2000)
Craver, C.F.: Explaining the Brain: Mechanisms and the Mosaic Unity of Neuroscience. Oxford University Press, Oxford (2007)
Craver, C.F., Kaiser, M.I.: Mechanisms and laws: clarifying the debate. In: Chao, H., Chen, S., Millstein, R.L. (eds.) Mechanism and Causality in Biology and Economics, pp. 125–146. Springer, Dordrecht (2013)
Cummins, R.: "How does it work" versus "what are the laws?": Two conceptions of psychological explanation. In: Keil, F., Wilson, R.A. (eds.) Explanation and Cognition, pp. 117–145. MIT Press, Cambridge, MA (2000)
Dale, R.: The Possibility of a Pluralist Cognitive Science. J. Exp. Theoret. Artif. Intell. **20**(3), 155–179 (2008)

Dale, R., Dietrich, E., Chemero, A.: Explanatory pluralism in cognitive science. Cogn. Sci. **33**, 739–742 (2009)
Fazekas, P., Kertész, G.: Causation at different levels: tracking the commitments of mechanistic explanations. Biol. Philos. **26**, 365–383 (2011)
Gardner, H.: The Mind's New Science: A History of the Cognitive Revolution. Basic Books, New York (1985)
Glennan, S.: Mechanisms and the nature of causation. Erkenntnis **44**, 49–71 (1996)
Glennan, S.: Rethinking mechanistic explanation. Philos. Sci. **69**(3), S342–S353 (2002)
Glennan, S.: The New Mechanical Philosophy. Oxford University Press, Oxford (2017)
Goldstein, B.: Cognitive Psychology: Connecting Mind, Research and Everyday Experience, 4th edn. Cengage Learning, Stamford, CT (2015)
Green, C.D.: Why psychology isn't unified, and probably never will be. Rev. Gen. Psychol. **19**(3), 207–214 (2015)
Harnish, R.M.: Minds, Brains, Computers: An Historical Introduction to the Foundations of Cognitive Science. Blackwell Publishers, Malden (2002)
Kant, I.: Critique of pure reason. In: Guyer, P., Wood, A., (eds.), The Cambridge Edition of the Works of Immanuel Kant. Cambridge University Press, Cambridge. (1998, Originally published in 1781)
Kaplan, D.M.: Explanation and description in computational neuroscience. Synthese **183**(3), 339–373 (2011)
Kaplan, D.M. (ed.): Explanation and Integration in Mind and Brain Science. Oxford University Press, Oxford (2017)
Kim, J.: Philosophy of Mind, 3rd edn. Westview Press, Boulder (2011)
Krickel, B.: The Mechanical World: The Metaphysical Commitments of the New Mechanistic Approach. Springer (2018). https://doi.org/10.1007/978-3-030-03629-4
Leahey, T.H.: A History of Psychology, 4th edn. Prentice-Hall, Upper Saddle River, NJ (1997)
Machamer, P., Darden, L., Craver, C.: Thinking about mechanisms. Philos. Sci. **67**, 1–25 (2000)
Milkowski, M.: Explaining the Computational Mind. MIT Press, Cambridge, MA (2013)
Milkowski, M.: Integrating cognitive (neuro)science using mechanisms. Avant: J. Philos.-Interdisc. Vanguard **2**, 45–67 (2016)
Milkowski, M. et al.: From wide cognition to mechanisms: A silent revolution. Frontiers in Psychol. **9**, (2018). https://doi.org/10.3389/fpsyg.2018.02393
Milkowski, M., Hohol, M., Nowakowski, P.: Mechanisms in psychology: The road towards unity? Theor. Psychol. **29**(5), 567–578 (2019)
Miller, G.A.: The cognitive revolution: A historical perspective. Trends Cogn. Sci. **7**(3), 141–144 (2003)
Newell, A.: Unified Theories of Cognition. Harvard University Press, Cambridge (1990)
Ochsner, K.N., Kosslyn, S.M. (eds.): The Oxford Handbook of Cognitive Neuroscience, vol. 1. Oxford University Press, Oxford (2014)
Piccinini, G.: Computing mechanisms. Philos. Sci. **74**(4), 501–526 (2007)
Piccinini, G.: Computationalism. In: Margolis, E., Samuels, R., Stich, S. (eds.) The Oxford Handbook of Philosophy of Cognitive Science, pp. 222–249. Oxford University Press, New York (2012)
Piccinini, G.: Physical Computation: A Mechanistic Account. Oxford University Press, Oxford (2015)
Piccinini, G., Bahar, S.: Neural computation and the computational theory of cognition. Cogn. Sci. **37**(3), 453–488 (2013)
Piccinini, G., Craver, C.: Integrating psychology and neuroscience: Functional analyses as mechanism sketches. Synthese **183**(3), 283–311 (2011)
Reisberg, D. (ed.): The Oxford Handbook of Cognitive Psychology. Oxford University Press, Oxford (2013)

Samuels, R., Margolis, E., Stich, S.: Introduction: Philosophy and cognitive science. In: Margolis, E., Samuels, R., Stich, S. (eds.) The Oxford Handbook of Philosophy of Cognitive Science, pp. 3–18. Oxford University Press, New York (2012)
Smith, E.E., Kosslyn, S.M.: Cognitive Psychology: Mind and Brain, International edn. Pearson, Edinburgh (2014)
Stephan, A., Walter, S. (Hrg.).: *Handbuch Kognitionswissenschaft*. Metzler, Stuttgart (2013)
Sullivan, J.A.: Construct stabilization and the unity of the mind-brain sciences. Philos. Sci. **83**, 662–673 (2016)
Teo, T.: Ontology and scientific explanation: Pluralism as an a priori condition of psychology. New Ideas Psychol. **28**, 235–243 (2010)
Thagard, P.: Hot Thought: Mechanisms and Applications of Emotional Cognition. MIT Press, Cambridge, MA (2006)
Thagard, P.: Why cognitive science needs philosophy and vice versa. Top. Cogn. Sci. **1**, 237–254 (2009)
Thagard, P.: Cognitive science. In: E.N. Zalta (ed.) The Stanford Encyclopedia of Philosophy, Winter 2018 edn. https://plato.stanford.edu/archives/win2018/entries/cognitive-science/ (2018)
Thagard, P.: Brain-Mind: From Neurons to Consciousness and Creativity (Treatise on Mind and Society). Oxford University Press, Oxford (2019)
Von Eckard, B.: What Is Cognitive Science?. MIT Press, Cambridge, MA (1993)
Walter, S.: Kognition. Reclam, Stuttgart (2014)
Ward, J.: The Student's Guide to Cognitive Neuroscience, 3rd edn. Psychology Press, New York (2015)
Weiskopf, D.: Models and mechanisms in psychological explanation. Synthese **183**, 313–338 (2011)
Weiskopf, D., Adams, F.: An Introduction to the Philosophy of Psychology. Cambridge University Press, Cambridge (2015)
Woodward, J.: What is a mechanism? a counterfactual account. Philos. Sci. **69**, S366–S377 (2002)
Wright, C., Bechtel, W.: Mechanisms and psychological explanation. In: Thagard, Paul (ed.) Philosophy of Psychology and Cognitive Science, pp. 31–79. North Holland Elsevier, Amsterdam (2007)
Zednik, C.: Mechanisms in cognitive science. In: Glennan, S., Illari, P. (eds.) The Routledge Handbook of Mechanisms and Mechanical Philosophy, pp. 389–400. Routledge, London (2018)

Part I
The Twenty-First Century Mechanistic Theory of Human Cognition

Chapter 2
The Mechanistic Theory of Complex Biological Systems

2.1 Central Theoretical Components of MTCBS

2.1.1 Brief Introduction to the Major Historical and Philosophical Foundations

The usage of 'mechanistic notions' to formulate explanations of natural phenomena has a very long history, which leads back at least to the work of the Ancient Greek pre-Socratic philosophers Democritus (c. 460–c. 370 BC) and Leucippus (fl. fifth centaury BC) (Bechtel 2008, p. 10) and the ancient philosophers Epicurus (341–270 BC) and Lucretius (99 BC–c. 55 BC) (Thagard 2006, p. 4). All these philosophers tried to explain, in one way or the other, the natural phenomena in terms of their constitutive parts, their motions, properties and interactions. However, these notions and the term 'mechanism' itself are used in a variety of senses through the centuries and it is not clear whether there is a central single scientific or philosophical systematic doctrine that encompasses all its historical uses. Craver (2007, p. 3) points out that authors who have been called mechanical philosophers in the literature differ from one another in many particular aspects. Moreover, it is not clear whether these specific historical uses have any current significant application in science or philosophy.

At any rate, the French philosopher, scientist and mathematician René Descartes (1596–1650) is considered to have most prominently played a role in the development of the early modern mechanistic philosophy as well as shaped mechanistic explanations for subsequent centuries. According to Bechtel, Descartes was one of the main advocates of the new mechanical science of the seventeenth century. The central idea was to replace the Aristotelian framework by explaining all phenomena of the natural world in terms of mechanical processes (Bechtel 2008, p. 1). Included in this Cartesian view, Bechtel continues, was not only ordinary physical processes but also processes in living organisms, i.e. all animal behavior and all human behavior comparable to that of other animals (that is, not language and reasoning) were mechanically explained (Bechtel 2008, p. 1; cf. Wright and Bechtel, 2007). At a certain point, the

D. A. Leite, *The Twenty-First Century Mechanistic Theory of Human Cognition*, Cognitive Systems Monographs 41, https://doi.org/10.1007/978-3-030-63680-7_2

mechanistic explanatory strategy became less attractive above all in the domain of physics of that period due to the work of Isaac Newton (1643–1727) and the popularity it achieved. It is true that Newton advocated what is now called a physics of 'mechanics', or *classical mechanics*, but this, in the view of the twenty-first century mechanists, is different from the mechanistic philosophy defended by traditional mechanist philosophers such as Descartes. Contrarily to genuine mechanists, which attempt to explain how things work by accounting for their physical parts and organization, Newton emphasized the role of general laws in explanations, such as his three laws of motion, which could be applied to an enormous variety of natural phenomena, including those who were explained by the mechanistic point of view of the time (Bechtel 2008, p. 11).[1]

In the seventeenth century, the framework of Newton became the most promising for the unification of natural philosophy. Nevertheless, Descartes' mechanistic approach remained influential especially in biology. As is well known, he used frequently in his explanations metaphors with human-made machines in order to propose explanations of biological systems. For instance, clocks and mills, which could move on their own, were used to explain the motion of organisms, and the hydraulically moved statuary in the Royal Gardens were appealed in order to provide a model of certain activities of the nervous system (Bechtel 2008, p. 12).

In the eighteenth century, the influence of Descartes' mechanistic approach was felt all over. One of the most important authors of this period to mention in this context is the French physician and philosopher Julien Offray de La Mettrie (1709–1751), which is considered to be one of the earliest French materialists of the eighteenth-century Enlightenment. In his works, as e.g. *Histoire Naturelle de l'Âme* (1745), or *L'Homme Machine* (1748), he expresses his mechanistic materialist views. Mental processes according to him were generated by the brain, a highly complex structure composed by physical parts, and not by an immaterial substance. Human beings should be considered machines, composed by physical parts as any other material object in the world. Since for him there is no abrupt difference between other animals and humans, which are just more complex, La Mettrie is considered to have expanded to humans the mechanistic framework that was part of Descartes philosophy of biology.

The mechanistic approach in the nineteenth century continued to be influential. Important scientific figures embraced the view, according to Bechtel (2008, p. 12). And there were significant advancements in consolidating some of the most fundamental mechanistic ideas. One factor that strongly contributed to this was the great controversy in the field of biology over the question of how to explain 'life' (cf. McLaughlin 1992; Stephan 1992). The issue concerned whether life should (or should not) be understood in a simpler physio-chemical way. On one side of the debate, there

[1] However, Thagard, one of the major proponents of mechanistic explanations in cognitive science, considers "Newton's theory of motion" together with the "atomic theory of matter" some of the "triumphs" of "mechanistic explanations", which has been "fabulously successful in modern science" (2006, p. 5). One can note, therefore, that the historical account of mechanistic explanations and what belongs or not to it can vary considerably from one influential neo-mechanist to the other. On the contributions of Newton to mechanistic explanations their interpretations sharply differ.

were the *mechanists*. In their view, the properties of living organisms could be fully explained in terms of the properties and relations of its physio-chemical parts. On the other side, there were the *vitalists*. In their view, organic matter differed fundamentally from inorganic matter and what accounted for the properties of living organisms was not the arrangement of their constitutive physical and chemical parts, but a primitive substance, called 'entelechy', which is embodied in the organism and guides the vital processes. By the end of the century, the mechanistic view was winning the debate and becoming increasingly more influential. Vitalism was dismissed from biology almost overall, but the debate was so intense that it continued with some force until the beginning of the next century.

In the middle of the twentieth century some important philosophers of science started to pay more attention to the significant role mechanistic explanations were playing concerning the debates about the origin of life. These philosophers attempted then to give a clear and detailed account of scientific mechanistic explanations. One prominent example is the American philosopher Ernest Nagel (1901–1985), in his famous work *The Structure of Science* (1961). This account strongly influenced the mainstream understanding of mechanistic explanation in philosophy of science until more recent times. However, Nagel's discussions about scientific explanations and the relations between theories in science and his discussion concerning mechanistic explanations were based primarily in examples taken from physics, which was considered highly advanced and at the position to provide normative scientific guidelines to the rest of the sciences. Moreover, Nagel's views on mechanistic explanations were influenced by his background in the philosophy of logical empiricism, dominant at the time.

Nevertheless, some contemporary philosophers set to themselves the task of providing a clearer and more well-articulated mechanistic approach to scientific explanation. This new turn of philosophy (but especially of philosophy of science) in the direction of establishing foundations for the mechanistic theory of scientific explanation started to take place in the second half of the twentieth century. Based on such developments, a more systematic contemporary formulation of, what I call, the *Mechanistic Theory of Complex Biological Systems* (MTCBS) began to be more clearly articulated in the decade of 1990,[2] when some philosophers started to analyze what biologists mean by mechanism and how they use mechanisms to explain particular biological phenomena (Bechtel 2012, p. 47). The philosophers responsible for developing this new framework were also practitioners of history of science, but,

[2]The first work articulating more systematically the new mechanistic theory of scientific explanation was the work of William Bechtel and Robert Richardson: *Discovering Complexity* (first published in 1993). Just three years after, Stuart Glennan, in his paper *Mechanisms and the Nature of Causation* (1996), started to develop his work linking causation and mechanisms. The paper of Peter Machamer, Lindley Darden and Carl Craver, *Thinking about Mechanisms* (2000), synthetized these works and elaborated these ideas (Craver and Tabery 2015, Sect. 1). All these works were central for developing the most fundamental ideas behind the new mechanistic approach. For the earliest developments of the contemporary mechanistic framework of scientific explanation were very important the works of Herbert Simon, *The Architecture of Complexity* (1962), Stuart Kauffman, *Articulation of Parts Explanations in Biology and the Rational Search for Them* (1971), and William Wimsatt, *Reductive Explanation: A Functional Account* (1976).

contrarily to the logical empiricists, their focus was on the biological, rather than physical, sciences (Craver and Tabery 2015, Sect. 1). For these philosophers, the kind of mechanistic explanation advanced in biology was the correct approach to biological phenomena and other phenomena investigated by the life sciences, rather than the explanatory framework advanced by the logical empiricists.[3] This mechanistic framework is considered by its proponents to be the first adequate analysis of what mechanisms are and how they work in science (Machamer et al. 2000, p. 2). Recently, MTCBS has been articulated in more detail and discussed in several works by several authors (cf. e.g. Bechtel and Abrahamsen 2005; Bechtel 2008; Craver 2007; Craver and Tabery 2015; Glennan 2017; Krickel 2018).[4]

2.1.2 Mechanism: The Central Concept Under Analysis

The technical use of the term 'mechanism' by leading neo-mechanists in order to characterize their most important concept varies significantly in the specialized literature. However, there are also important common elements. A 'mechanism', understood as a biological system, can be characterized as "a structure performing a function in virtue of its component parts, component operations, and their organization" (Bechtel 2008, p. 13; cf. Bechtel and Abrahamsen 2005; Craver 2007; Glennan 2017). To put it in another way: "A mechanism is simply a composite system organized in such a way that the coordinated operations of the component parts constitute the mechanistic activity identified with the explanandum" (Bechtel and Wright 2009, p. 119). Accordingly, the core idea is that mechanisms are systems made of component parts, their operations and their general organization. The component parts perform operations and interact causally with other parts of the mechanism, producing a certain phenomenon. The general behavior of the whole system is a result of the specific organization of the components and their interactions. Examples of mechanisms in neuroscience are neurons that fire, neurotransmitters that bind to receptors, brain regions that process information, and mice that navigate their environments (Craver 2002, p. 84). A mechanistic explanation should describe how the organized functioning of the system is responsible for generating the phenomenon being investigated.

---

[3] I will discuss this point more deeply in this chapter, Sect. 2.1.4.

[4] The history of what has been considered 'mechanistic explanation' in science and philosophy is long and complex. Evidently, it is not my purpose to provide a comprehensive and exhaustive historical view here. The purpose is rather: (1) to give an idea of how complex this history is, which can involve many authors holding complex views on the issue in various debates in different contexts along the centuries (these views would have to be properly contextualized, analysed and related in a comprehensive historical account that can be offered elsewhere); and (2) to contextualize the particular formulation of mechanistic explanation in its twenty-first century version – the only one that is my goal to critically analyse and evaluate. For more detailed discussions of historical aspects of the framework, cf. Glennan and Illari (2018, Part I).

The phenomenon is an occurrence in the world and is what needs to be explained, i.e. the *explanandum*. Typically, this is some behavior of the entire mechanism (Craver 2007, pp. 139, 212). It is also described functionally and this function can be performed since the mechanism as a whole has some particular properties and capacities. The neo-mechanists use the concept of 'behavior' in a very loose way, though. A non-natural mechanism, such as a clock, a car, a motorbike, or an airplane, produces 'behaviors'. In the same sense, cells, hearts, stomachs and neural systems of animals produce 'behaviors'. And in the same sense, an animal, such as a mouse, a horse, or a human, produce a 'behavior' when this animal, for example, runs because it is afraid of a noise, or moves around looking for food. Given the ambiguity in the concept of 'behavior', many authors use the concept of a 'function' in the biological sense, which can be perhaps less controversial. Thus, a biological mechanism functions, or performs a function, i.e. it is a biological functional mechanism (cf. Piccinini 2012, p. 230). In mechanistic explanations the phenomenon needs to be characterized in the most precise way possible and all its aspects should be also described with the highest possible level of accuracy. The mechanism is the whole system that is 'responsible' for the occurrence of the phenomenon under consideration, which is determined by relevant component parts, operations and organization—these are the *explanans*, i.e. what is required for the explanation (Craver 2007, p. 144).

Mechanisms, thus, have essentially three central constitutive elements. The first one can be called *component parts* (or entities, or structures).[5] These are not just simply parts, but *working parts* of a given mechanism. One could break a mechanism in many different parts that do not work properly anymore. But one should decompose a mechanism in its working parts to understand how they properly work. Thus, component parts can be understood as pieces that make identifiable contributions to the function of the mechanism (Craver 2007, p. 188). The working parts refer to the structural component of the mechanism (Bechtel 2008, p. 14). They are in space and time and are the kinds of things that have, sizes, masses, carry charges, and transmit momentum (Craver 2007, pp. 5, 6). In a great amount of mechanistically oriented work done, for example, in the fields of cognitive science and neuroscience, the research focus is on finding relevant mechanistic structures and structural components in order to explain cognition. Numerous works have been done at the level of molecules and neurons in brains, as well as at the level of smaller and larger neural networks.

The second element can be called *component operations*. Mechanisms and their working components have a set of particular properties (or capacities), and some of them allow these structural components to perform operations. The operation of a component part of a mechanism can be considered at the same time a *phenomenon*, i.e. a function relative to this component part taken itself as a mechanism (not as a sub-mechanism of a given mechanism), but just in the case that this function is what

[5]There are many disputes concerning terminology among proponents of the neo-mechanistic framework, not just referring to these three core elements, but also to other characterizations. I will be following generally Bechtel's terminology here, since his work is one of the most influential concerning cognitive science.

one wants to explain, i.e. the *explanandum*, not the *explanans*. Such operations can be considered as causal changes involving parts (Bechtel 2008, p. 14; cf. Craver 2007, pp. 5, 6); they allow these working parts to engage in a variety of causal relations. In this sense, these operations are productive (they produce other operations) and are the causal component inside and outside mechanisms. Accordingly, I can fix my usage of these two terms in the following manner: I will speak of 'parts' and 'component parts' when I am referring to the structure performing the operation; I will speak of 'operations' and 'component operations' when I am referring to the functions or activities of the structures. When it is irrelevant to distinguish between component parts and component operations or it is referred to both of them together, I will simply use the term 'component' (in this, I follow Bechtel and Abrahamsen 2005, p. 425).

The third central element constitutive of a mechanism is its *organization*. Component parts and component operations in mechanisms have a proper organization, i.e. a specific way in which they are related, or in which they interact to one another. The mechanistic form of organization is particular since biological mechanisms are not mere aggregates (in which the whole is literally the sum of its parts). Mechanisms are more than the sum of their parts, and it is because of their organization that they are able to perform functions that their parts cannot do individually (Craver 2007, p. 227). In the case of the neural system, the organization of mechanisms involves, for instance, spatial relations between brain areas and temporal ordering of their operations. The organization of a neural system is also provided by patterns of cellular connectivity through which neurons in different areas communicate with each other (Bechtel 2008, p. 119). Since on average each neuron is connected to a thousand other neurons, such patterns are enormously complex. The different forms in which the component parts and operations are related to each other is fundamental for the mechanism as a whole to function in a particular way. This is why biological neural mechanisms can be understood as systems that are organized in a specific way so that their operations produce the particular phenomena that need to be explained.

Different kinds of organization are present in different mechanisms. For instance, there is a simple linear organization present in systems with a lower degree of complexity, such as bikes, clocks, cars and airplanes. However, in more complex systems, such as living organisms, another kind of organization is found. In these living systems more complex modes of organization, like cyclic pathways and (positive and negative) feedback loops, are present and are critical for the general behavior of the organism (Bechtel 2008, p. 17).

Moreover, for the understanding of the behavior of the entire mechanism it is also important to consider the *environmental conditions* in which this mechanism produces its activities. In fact, as Bechtel points out, the behavior of mechanisms is highly dependent on conditions in their environment (2009b, p. 559). For example, environmental factors can exercise a great impact on how humans perceive visual stimuli. This means that, in order to understand the behavior of a biological mechanism, it is necessary to consider not just its internal working parts (and their respective operations) and its internal complex organization, but also the particular kind of the inputs it receives from the environment (Bechtel 2009b, p. 557).

In a nutshell, the whole mechanism is a physical complex structure, composed of working parts that perform particular operations. These component parts and component operations are organized in a specific way. This internal organization together with external conditions provide the whole mechanism with the capacity of performing physical functions (a neural mechanism, for instance, performs neural functions and in some cases also cognitive functions), giving rise thereby to a phenomenon which is the object of a scientific mechanistic explanation.

Let us consider more formally, for the sake of clarity and objectivity, all the important elements in this theory of mechanisms and how they are related. Let us start with a given whole, *W*. This whole is a biological mechanism (a biological system) that has a particular physical structure, S_p, and that performs a set of physical functions, F_1, F_2, F_3, F_4, each of which has its particular functional configuration, S_f. A general mechanistic explanation aims to explain the particular functions, F_1, F_2, F_3, F_4 of *W*, which are ultimately the phenomena to be explained, i.e. the *explanandum*. Let us assume that *W* is a neural mechanism, that F_2, F_3 and F_4 are merely particular neuro-physiological processes without any relevance for any cognitive process, and that F_1 is rather a neuro-physiological process that somehow 'produce' a particular cognitive function.

Now, to explain this particular cognitive function, F_1, of the neural mechanism *W*, it is important to consider the particular physical structure, S_p, of *W*, and S_f, the functional configuration of F_1 performed by *W*. S_p is composed by the physical structural component parts, C_1, C_2, C_3, C_4 and by their internal physical organization, O_c. S_f, in turn, is composed by particular physical sub-functions (activities performed by the physical structural component parts of *W*), A_1, A_2, A_3, A_4 and their internal organization O_a. Finally, it is also important to consider the causal influence of external factors in the physical structural organization E_c, and in the physical functional organization, E_a. Thus, the basic elements of a mechanistic explanation of a cognitive function are as follows. *W* produces a particular cognitive function F_1: $W \rightarrow F_1$. This phenomenon is the target of the explanation. This explanation is achieved through the consideration of the components of *W* and the external elements causally affecting the production of its function. *W* has a S_p and performs F_1, which has a S_f. But $S_p = C_1, C_2, C_3, C_4 + O_c$, and $S_f = A_1, A_2, A_3, A_4 + O_a$. Hence: $W \rightarrow F_1 = S_p(C_1, C_2, C_3, C_4 + O_c) + S_f(A_1, A_2, A_3, A_4 + O_a) + E_c + E_a$.

2.1.3 A Theory of Scientific Explanation

Science is an activity performed by human beings and one of its objectives is to describe the world we live in in the most systematic and accurate way. These descriptions are constantly revised and improved so that science can provide us with the currently best possible description of what is happening in nature. But science aims at much more than solely describing phenomena in the world. One of its foremost objectives is to provide knowledge about the reality we live in, i.e. to explain the phenomena in the world and to answer the question 'why?' rather than only the question 'what?' (Hempel and Oppenheim 1948/1965, p. 245).

Thus, science has the purpose to provide explanations, not merely descriptions, for why things in nature happen in the way they do. It assumes that phenomena in the world occur not arbitrarily, accidentally or spontaneously, but rather because there is a certain pattern of regularity that accounts for their occurrence and the way they occur. Scientific activity, thus, uses these regular patters in order to provide explanations for the occurrence of the natural phenomena. For example, why do physical bodies fall? Why planets are in motion? Why do some animals present a certain pattern of behavior? How does the human brain work? Scientific explanations of phenomena such as these are also provided through careful and systematic observation, experimentation, logical and mathematical reasoning, and other activities that help in the development of particular and comprehensive theories about how and why things happen. With such explanations and theories, phenomena in our immediate environment and in the world can be predicted, manipulated and controlled, all for our own benefit.

However, there is a great amount of controversy concerning what exactly scientific explanations are and how exactly they should be constructed. Given the diversity found in science and the degree of expertise in present days, it is also considerably disputed whether it is really possible to formulate a theory of scientific explanation that could be applied to all different sciences. For a very long-time scientists and philosophers have been discussing such matters (cf. Woodward 2014).

Currently, one of the theories of scientific explanation most advocated is the *Mechanistic Theory of Scientific Explanations in Biological Science* (MTSEBS)—which is part of MTCBS. According to its proponents, MTSEBS can provide a new framework for addressing many traditional philosophical issues (Machamer et al. 2000, p. 1). It is also described as a new mechanistic philosophy of science that provides different answers to long-standing questions and normative prescriptions for many sciences, including cognitive science (cf. Bechtel 2009a, p. 549). The leading idea behind mechanistic scientific explanations is that it is mainly applied to the biological sciences (its general scope is still controversial). Roughly, the theory states that a satisfactory scientific explanation in the biological sciences requires providing a description of a biological mechanism. This means that every biological system and their functions (including cells, organisms, the phenomenon of life, and cognitive capacities) should be explained in terms of complex physical-biological mechanisms. Accordingly, MTCBS and MTSEBS are the underlying theories that provide philosophical and theoretical foundations to the mechanistic theory of human cognition, present in contemporary cognitive science and cognitive neuroscience.

2.1.4 A More Plausible Theory of Scientific Explanation for the Biological Sciences

MTSEBS is articulated in strong opposition to the *Deductive-Nomological Theory of Scientific Explanation* (DNTSE) defended most prominently by Carl Hempel (1905–1997) and overwhelmingly accepted in the mid-twentieth century (cf. Hempel and Oppenheim 1948/1965; Hempel 1965; Nagel 1961). This theory is one of the most important and well-articulated theories of scientific explanation ever proposed and is still highly influential. It states that a scientific explanation of a given phenomenon has two major components: the phenomenon that is being explained (*explanandum*); and what provides the explanation (*explanans*). Both these components are formulated in form of sentences that are supposed to account for the explanation of the phenomenon (Hempel and Oppenheim 1948/1965, p. 247). Moreover, in a proper scientific explanation the phenomenon being explained needs also to be logically deduced from the statements explaining it, as in a common process of logical deductive reasoning. According to this theory, thus, explanations are achieved by a proper deduction of the *explanandum* from the *explanans*. In this way, the scientific explanation takes the form of a deductive argument (Woodward 2014, Sect. 2).

According to DNTSE, in order to explain an event, it is necessary to show that it can be deduced from a law of nature (or a set of laws) and initial particular conditions. Thus, in the sentences that constitute the premises of an argument formulated in order to explain a given phenomenon there need to be at least one 'law of nature', which plays a substantial role in the derivation of the conclusion. As Hempel and Oppenheim state, "the explanation of a general regularity consists in subsuming it under another, more comprehensive regularity, under a more general law" (1948/1965, p. 247). The authors give the example of "Galileo's law for the free fall of bodies near the earth's surface" which "can be explained by deducing it from a more comprehensive set of laws, namely Newton's laws of motion and his law of gravitation, together with some statements about particular facts, namely, about the mass and the radius of the earth." (Hempel and Oppenheim 1948/1965, p. 247). Another example, given by Woodward (2014, Sect. 2), is the following. From Newton's laws of motion, the inverse square law governing gravity, together with information about the mass of the sun, the mass of Mars and the present position and velocity of each (initial conditions), it is possible to explain the motion of Mars, and also to predict the future position of the planet. This is a case of a scientific explanation of the motion of a planet achieved through a deductive argument where Newton's laws play a central role, together with initial conditions.

Accordingly, theories in this framework explain particular events happening in a particular space and time constructing general regularities or laws (roughly, general repetitive and regular patterns of connection found between phenomena in nature). This is related to the common idea that laws of nature account for events in the natural world (that the occurrence of natural events follow regular patterns, described through inviolable general laws), indicating what must happen if certain conditions obtain. A given scientific framework ultimately aims, according to DNTSE, at discovering such laws—regularities of broad or even universal applicability, such as Newton's laws of motion or the laws of thermodynamics.

As one can note, this framework is greatly influenced by physics, which in many moments of history until present days has been considered by many authors the highest exemplar of a genuine empirical science. The main examples for the construction of the theory are indeed taken from the physics of that period. However, even if physics plays a major role in the construction of this theory of scientific explanation, the general principles of the framework apply equally to chemistry, biology, psychology, sociology, economy, history, linguistics, and indeed to all other sciences (Hempel and Oppenheim 1948/1965, p. 251). It is, thus, a very comprehensive theory of scientific explanation.

Even though DNTSE was meant to be a very comprehensive theory of scientific explanation, for the twenty-first century mechanists, it cannot be applied to all kinds of sciences with equal success, and in particular it is inadequate as a theory of explanation for the biological sciences (cf. Bechtel 2008, chap. 1; Craver 2007, chap. 2). Contemporary scientists working in the life sciences, such as biology, neuroscience and cognitive science, most frequently do not formulate explanations in terms of general laws, as it was believed by Hempel and others. Instead, they formulate explanations in terms of 'mechanisms' (Bechtel and Abrahamsen 2005; Machamer et al. 2000), i.e. successful explanations in the life sciences take the form of providing a description of how a given biological mechanism works.[6] Bechtel and Abrahamsen claim that in reading the literature on the biological science one gets convinced that 'mechanism' is indeed the term biologists most frequently invoke in explanations (2005, p. 422). Similarly, Craver points out that mechanistic explanations are to be found everywhere in neuroscience (2007, p. viii).[7] Thus, he concludes that in order to understand contemporary neuroscience, one needs to understand this type of explanatory framework (Craver 2007, p. viii).

It might well be true that mechanistic explanations are being used almost everywhere in biology, neuroscience and beyond—it is indeed important to understand them. However, this particular kind of argument is not solid as a way of providing support in favor of MTSEBS. The argument just shows that this kind of explanation is currently popular and fashionable and, therefore, it deserves the attention of someone seeking to understand the biological and neuroscientific way of constructing scientific explanations. It does not show, however, why they are better than other kinds of scientific explanations and why one should adopt them. After all, DNTSE is still considered by some authors to be superior as a theory of scientific explanations across the sciences. DNTSE has many positive aspects. For instance, this framework has been considered successful regarding explanations in physics, as we saw above, and physics has been considered for quite a long time to be the highest exemplar of

[6]There are also other alternatives for substituting the traditional theory of scientific explanation (cf. Woodward 2014). However, the neo-mechanists contend that their MTSEBS is superior to all other views.

[7]For Craver, cognitive psychology, experimental psychology, behavioral psychology and cognitive neuroscience are part of neuroscience; thus, whenever he talks about neuroscience in general, he is also talking about these sciences (cf. 2007, p. 16, 176).

a successful scientific field. Moreover, DNTSE has a very broad application and its principles are equally applied to all explanatory sciences. In this way, the theory can be viewed as integrating the sciences and their general scientific project according to a common scientific goal, i.e. the goal of providing scientific explanations of a particular kind in a particular way. Therefore, if one intends to argue for the superiority of another view, one needs to show why and/or where exactly DNTSE fails and the alternative view is successful.

A better way to argue for the superiority of MTSEBS for the life sciences is as follows. In the mechanistic view, logical empiricists got it wrong because they had physics as the primary and most important exemplar of science (Craver and Tabery 2015, Sect. 1). The most crucial point is that the search for mechanisms in the life sciences is different from the search for laws of nature in physics. Research aimed at finding mechanisms is usually more specific, since a mechanism consists of a particular set of parts that carry particular operations, organized so as to produce a given phenomenon. Thus, in the life sciences, researchers focus on particular aspects of the composition and organization of a mechanism (Bechtel 2008, p. 4). Generalizations in the biological sciences, such as neuroscience for example, are fragile, variable, and historically contingent. But the same does not happen with the gas laws or the laws of optics, for instance. The notion of strict/universal laws has, thus, little application in the domain of the biological sciences (cf. Craver 2007, p. 233–234; Craver and Kaiser 2013, p. 127). This makes such generalizations to be very different from the laws in physics.

What the neo-mechanists are normally saying is not that all scientific explanations are explanations of mechanisms; not even that absolutely all explanations in the life sciences, e.g. in biology, take the form of identifying mechanisms. The claim is that in the overwhelming majority of life sciences one *cannot* construct general laws to explain biological phenomena.

The analysis of biology and biological explanations show, in their view, that one looks in vain for laws—examples of laws in biology, such as Mendel's laws, have frequently turned out to be incorrect or at best only approximately correct (Bechtel 2008, p. 10).[8] Since biologists as well as psychologists rarely construct laws in their explanations (Bechtel 2009b, p. 544) one can argue that in most explanations in the life sciences and cognitive science researchers do not appeal to laws in the manner specified by the DNTSE (Bechtel and Abrahamsen 2005, p. 422). Normally, biologists explain phenomena such as cell respiration, embryological development, disease transmission, protein synthesis by identifying the components, operations, and organization of the related mechanisms in organisms (Bechtel 2008, p. 1). What is relevant here is not any law of motion, but rather, for example, questions such as

[8]Occasionally, though, some neo-mechanists put the point in a very modest way. They claim that 'in many cases' or 'frequently' explanations in the life sciences aim to describe biological mechanisms (Bechtel and Abrahamsen 2005, p. 422; Bechtel 2007, p. 174). This very modest way of expressing this idea appears to be in conflict with what they really want to offer, i.e. a new interpretation of how scientific explanations are constructed in the overwhelming majority of the biological sciences. Thus, this can be viewed as an occasional lack of clarity, or imprecision, in the way they formulate their idea.

the following. How the circulatory system in a dog, a cat, a horse or a human being works? What are the component parts of the circulatory system in a human being, what are the functions of these parts, and how are they connected? To which level of detail can we generalize the functioning of the human heart to all human beings? What happens if we have a problem in the human heart caused by a particular kind of external factor? How can we make the human heart to function in the best possible way?

Accordingly, in physics generalizations of scientific theories are much stronger than in biology. This is why in physics it is possible to construct general laws, i.e. universal conditional statements that apply to any situation in which the antecedent of the conditional is satisfied (Bechtel and Abrahamsen 2005, p. 437). The example given by the authors is Newton's first law of motion: If there is no force on a body, its momentum will remain constant. This law applies to any physical body in nature, without exception—it also applies to all physical bodies that existed in the past and all that could exist in the future. Certainly, there are no such 'general laws' with such degree of generality in biology, or in other life sciences, for that matter. In such sciences, models of mechanisms are developed for specific exemplars and are not represented in terms of universally quantified statements (Bechtel and Abrahamsen 2005, p. 421).

2.1.5 The Theory Applies to Neuroscience, Cognitive Neuroscience and Cognitive Science

In neuroscience, cognitive neuroscience and cognitive science, according to the proponents of the mechanistic framework, one can observe the same kind of explanation, i.e. the vast majority of explanations in these fields are not presented in terms of laws and logical derivations, but rather in terms of regularities discovered in mechanistic systems of organisms and individuals. As the current literature in cognitive science and neuroscience clearly indicates, the term 'mechanism' is frequently used when psychologists and cognitive neuroscientists offer explanations for mental activities (Bechtel 2008, p. ix, cf. p. 22; cf. Craver 2007, p. vii). Even if psychologists sometimes use laws in their explanations, such as Weber's law, they are not used in order to explain the phenomena, but rather to merely account for the regularities that require explanation (Bechtel 2007, p. 172). Thus, according to this view, when an explanation is offered in cognitive science, it is frequently in terms of cognitive mechanisms, just like in biology.

Bechtel claims that dividing the mind/brain into component systems has been a common strategy in psychological theorizing (2009c, p. 157). In his view, cognitive psychology attempts to identify different mechanisms that are responsible for different psychological capacities, e.g. different kinds of memory and different memory systems (Bechtel 2009c, p. 157). Similarly, according to Samuels et al., cognitive scientists seek to specify mechanistic explanations of psychological phenomena (2012, p. 5). Scholars working in this field overwhelmingly accept that the idea, generally associated with Descartes, concerning the existence of an immaterial soul, not subjected to physical laws, is implausible given scientific developments over the past three hundred years (Samuels et al. 2012, p. 5). Given this, cognitive science is an empirically informed approach that investigates issues about how mental capacities as perception, reasoning, and memory could result from activity of physical neural systems (Samuels et al. 2012, p. 5). In order to do this, the mechanistic framework has been widely used in cognitive science, cognitive neuroscience and in neuroscience. The brain as well as cognition here are understood in terms of mechanisms, being cognitive processes the informational processing operations done by particular brain components.

Most cognitive scientists and neuroscientists consider the brain to be composed of mechanisms (Craver 2007, p. 2) and assume that the mind is also a complex mechanism that produce cognitive phenomena (Bechtel 2008, p. 2). Indeed, Samuels et al. (2012, p. 10) clearly state that the "mechanistic assumption" is a foundational assumption in the philosophy of cognitive science, related to the widely held view that the mind is some kind of mechanism, that is, "a physical device decomposable into functionally specifiable subparts." Given this assumption, a central goal for cognitive science would be to describe and characterize the nature of this broad cognitive mechanism, and thus to provide an account of our cognitive architecture. In this view of Samuels et al. (2012, p. 10), a huge amount of cognitive science is ultimately concerned with the attainment of this goal.

The aim of neuroscience and cognitive science is, thus, according to this view, to account for the mechanisms that underlie and explain the brain and cognition. In this sense, in this neo-mechanistic view, all phenomena in cognitive science are to be explained in mechanistic terms, according to the mechanistic guidelines. Consequently, capacities such as perception, memory, attention, language, reasoning, and consciousness, and all the sub-capacities related to these general capacities are mechanistic in nature. Theories and concepts traditionally accepted in scientific psychology and overwhelmingly used, for instance, by many social psychologists, developmental psychologists, educational psychologists, cultural psychologists, clinical psychologists etc., need also to be accounted in mechanistic terms. Thus, complex capacities such as having complex beliefs, desires and emotions can and should equally be mechanistically explained. All aspects of human consciousness should also be mechanistically explained. As a result, all causal relations and all regularities among psychological capacities and functions (e.g. relations between different beliefs in a given belief system) and also between psychological functions, environment and behavior (e.g. relations between environmental objects of desires, desires, emotions, beliefs, intentions and actions) are best understood in terms of biological neural mechanisms and their simple or complex functions.

Ultimately, believing, desiring or feeling and all other cognitive phenomena should be understood as particular functions performed by certain highly complex neural mechanisms. And explaining the operations of these mechanisms is what is actually relevant in order to provide a genuine and successful scientific psychological explanation. Cognitive capacities are a kind of information processing that involves specific kinds of computations over specific kinds of representations ultimately taking place in complex neural mechanisms. Hence, no human cognitive capacity fall outside the mechanistic framework.[9]

In a nutshell, for the new mechanists, therefore, there is no necessity of logical derivation and general laws of nature for constructing a solid scientific explanation in the biological sciences, including cognitive neuroscience and cognitive science. Nor a scientific explanation needs to be constructed as a deductive valid argument, where the conclusion about what is being explained follows necessarily from the premises related to the content used to explain. Versions of the DNTSE in terms of inductive arguments and degrees of probability (cf. Woodward 2014, Sect. 2.3) are also incorrect ways of characterizing scientific explanations especially in the biological sciences, since such explanations are actually not constructed in terms of arguments where the conclusions simply follow from the premises. In the view of the proponents of MTSEBS, scientific explanations within the life sciences such as biology, neuroscience and cognitive science aim at understanding the works of biological mechanisms.

2.1.6 Mechanistic Scientific Explanations

The major goal of a mechanistic scientific explanation in, for instance, biology, cognitive neuroscience, and cognitive science, therefore, is to identify the component parts of a mechanism, investigate their operations and their organization, and finally, determine how these elements generate the target explanandum (Bechtel and Wright 2009, p. 120). The general idea is to make some sort of reverse engineering of the biological system under investigation (e.g. the visual system or the memory system of humans). In order to do this, the following steps are required: (1) a first *characterization of the phenomenon* (as accurate as possible); (2) the process of *decomposition*, which is divided in (a) *functional decomposition* and (b) *structural decomposition*; (3) the process of *localization*; (4) the *formulation of a mechanistic scientific theory (or model)* about how the system under investigation works.

Accordingly, the starting point of a general mechanistic explanation is usually a first general *characterization of the phenomenon* to be explained, with the highest degree of accuracy and detail as possible, together with a general characterization of the *putative particular mechanism* that is supposed to account for it. Once this is done, the process of *decomposition* can start. This is aimed at understanding how the component parts work, how their operations are performed and how they are organized in order to produce the phenomenon. Thus, to decompose a mechanism means to identify its working parts and their operations, separately. This mechanistic

[9]A detailed presentation of this mechanistic point of view will be made in Chap. 3, Sect. 3.1.2. A detailed discussion concerning the compatibility between cognitive/psychological explanations with mechanistic explanations will be carried in Chap. 6, Sect. 6.1.2, as well as a detailed discussion about the view of cognitive processes as neural information processing that involves a specific kind of 'computation' over a specific kind of 'representation', in Chap. 6, Sect. 6.1.3.

decomposition can be made in two ways: in structural terms through the decomposition of component parts (*structural decomposition*); and in functional terms through decomposition of component operations (*functional decomposition*).

After the decomposition is made on these two dimensions, it is also necessary to link specific component working parts to specific component operations. This is called *localization* (e.g. the suggestion that the brain structure called fusiform gyrus contains an area responsible for the function of recognizing faces). A further step is the understanding of the internal causal relations and interactions between the component parts and their component operations, i.e. how the mechanism as a whole is organized structurally and functionally. Finally, it is important to understand the role that external causal factors play in the performance of the mechanism under investigation considering its structural and functional dimension. These steps taken together make possible to construct a theoretical model about how the entire mechanism under consideration works.[10] Both the characterization of the phenomenon and the mechanism performing it can be revised through scientific development.[11]

The aim of a final general account of how the mechanism works is to reach a plausible account that explains how the organization of the component operations gives rise to the phenomenon investigated, which is produced by the whole mechanism. The proposed theory of a given mechanism determines the components that are relevant for explaining the phenomenon and leaves the components that are irrelevant out of the model.

Sometimes the explanation of a mechanism can be achieved simply by carefully observing how it works. But normally the task demands experimental procedures, manipulation and control, since frequently the complexity of some mechanisms does not reveal its functioning to an observer. Often, therefore, scientists must intervene trying to change the normal operations of a mechanism in order to understand better its structure, operations, organization and thereby overall functioning. Some of these interventions are used to delineate with more accuracy the phenomenon for which the mechanism is responsible; other interventions aim at understanding the internal operations of the mechanism under investigation.

There are two main ways of intervening on a mechanism in order to understand how it works: either by modifying its input or the external conditions under which it functions; or by modifying the internal operations of the mechanism itself. Consequently, a change can be detected at the level of the general function of the whole mechanism or at the level of the functions of its components. In behavioral experiments, made in neuroscience and cognitive neuroscience, for instance, the idea is to manipulate the task posed to the organism or to the subject in order to understand the mechanism that is responsible for that behavior. Thus, this kind of experiments

---

[10]The term 'model' here is being used in the technical sense of providing a scientific representation or simulation of a particular phenomenon or system in nature, its properties and how it happens or how it works.

[11]Some further aspects related with decomposition and localization in mechanistic scientific explanations will be discussed with more detail in Chap. 3, in the context of particular discussions concerning the mechanistic theory applied to human cognition, the human neuro-cognitive relationship, and the explanations in cognitive science.

do not offer a direct measure of the internal mechanistic structure itself or of any of its operations. In this case, sometimes the measures can be related to the accuracy with which the subject performs a specific task or slightly modified tasks in different conditions. Frequently, the measure is also in terms of the amount of time necessary to perform a particular task.

There are also experiments that manipulate directly internal operations inside the mechanism and measure the overall behavior resulting from the intervention or the change in the component's operation after the intervention.[12] According to Bechtel (2008, p. 41), two forms of internal intervention can be used: (1) impairment of the component part temporarily or permanently (e.g. lesion experiments); (2) stimulation of the component part making the operation it performs faster or more likely to happen (stimulation experiments).

Finally, there are experiments that measure internal physical operations in a mechanism working normally. For example, experiments and methodologies for the investigation of particular neural mechanisms (e.g. investigations of their location, electrical and chemical activity) and the attempt to relate them to cognitive operations. In the case of biological mechanisms such as brains, for instance, a common way to do research is to present a variety of stimuli to the different senses, such as tactile, visual, auditory, olfactory or gustatory stimuli, then to measure the electrical activity in the brain using recording from implanted electrodes. This recording can be done with single cells (single cell recording) or many cells (multi-cell recording). However, it is a very invasive method since the brain needs to be exposed and normally when the recording is made, the cells are destroyed or injured.

Due to these limitations, researchers frequently use another method for recording electrical activity in the brain, called electroencephalogram (EEG). Unfortunately, this technique is limited in its space resolution and it does not offer too precise information about from which areas are the data being recorded. To offer yet another methodological possibility, during the 1980s and 1990s two methodologies that give scientists a relatively high space resolution became available: positron emission tomography (PET) and functional magnetic resonance imaging (fMRI). Especially using the second technique, it is possible to obtain a relatively good spatial measurement of activities happening in different brain regions while a cognitive function (task) in being performed. This technique takes advantage of the correlation between increasing of certain metabolic processes in neurons and increasing of brain activities on that area.

Nonetheless, relating cognitive functions to brain activity is no simple task. To start with, PET and fMRI are not direct measures of brain activity an even the best of such kind of methodology currently present limitations in its temporal resolution. The results need to be analyzed in a very careful way: it is not just a matter of putting the subject inside the machine, discover what 'lights up' and relate the particular region

[12]However, in the case of humans, experimental psychologists need often to be very ingenuous to find evidence about how internal processes works without direct access to them, leaving also aside difficulties related to ethical limitations.

to the particular task being performed, as Bechtel correctly notes (2008, p. 47).[13] It is important to emphasize that all these techniques used for investigating mechanisms and for providing mechanistic explanations have their limitations and none of these methods alone can offer complete information or be enough to establish a strong and more general scientific theory with high plausibility. Together, however, they have more strength, are more robust and can be used to provide a very well-articulated and useful theory about how a particular neural mechanism works. This theory then can be constantly improved by scientific activity towards the ideal goal of understanding the mechanism completely in all its minor details.

2.1.7 Conditions for Assessing Mechanistic Explanations

According to Craver (2007), an account of mechanistic explanations should not be merely descriptive, i.e. merely describe how explanations are constructed in fields such as biology, neuroscience and cognitive science. Rather, it should prescribe norms of explanation as well and determine what counts as an acceptable mechanistic explanation (Craver 2007, p. 20). In his view, the mechanistic framework needs to provide a set of norms by which explanations should be assessed. The main issue, he points out, is to make the idea of mechanistic explanations precise and normatively rigorous (Craver 2007, p. 111). Bechtel agrees, and states that the mechanistic framework indeed advance normative prescriptions and provide normative guidance to cognitive science, thereby influencing it, and contributing to it, in the most direct way (2009a, pp. 564, 565). In other words, this means that the mechanistic theory provides criteria to: (1) establish when scientific explanations are mechanistic and when they are not; and (2) distinguish good mechanistic explanations from bad mechanistic explanations.

There are, accordingly, four important assessment conditions to be considered in order to evaluate a mechanistic explanation of a given phenomenon. The first condition is related to the fact that a mechanistic explanation can be understood as a continuum that starts with a conjecture about what model could possibly provide the explanation for the phenomenon under investigation and end with an actual model that explains how the function is actually performed in nature (cf. Craver 2007,

[13]The issue of localizing cognitive functions in particular areas of the brain remains very controversial in present days. For there is great variability between subjects in the scientific studies made with recent methodology, given the enormous complexity of certain cognitive capacities and processes. For a more detailed discussion on this cf. Bechtel (2002) and Uttal (2001, 2011, 2012). Moreover, the particular issues concerning multiple realizability (i.e. roughly, the possibility that cognitive functions are performed by a variety of neural structures) offer some difficulties for the thesis that particular neural structures are the only ones that perform particular cognitive functions. For a more detailed discussion of multiple realizability cf. Chap. 6, Sect. 6.1.3. Another related problem is that the more complex a system is, more difficult it is to decompose it in simpler parts and point out in detail what the particular functions for each part of the system are. Therefore, for the case of highly complex human cognitive functions and their putative neural substrate this strategy is troublesome. For a more detailed discussion of this point cf. Chap. 5, Sect. 5.1.1 and 5.1.2).

p. 112). Thus, this dimension is a matter of progress in terms of the plausibility concerning different possible mechanistic explanatory models of a given biological mechanism. To better characterize the development of the scientific explanation in this continuum, it is useful to divide it in at least three categories.

The first is *possible mechanistic models*. Mechanistic models in this category are only conjectures loosely constructed about the kind of mechanism that could produce the explanandum phenomenon. An example here would be the early work done by classical simulation of cognitive capacities by symbolic digital computers. Such simulations present a possible way in which certain component parts and operations of computers produce a given phenomenon that is very similar to the cognitive phenomenon under investigation in all the relevant respects, but they do not show that the natural way such phenomenon is produced by natural human cognition is exactly the same as described by such computational model. Such models are useful at the start because with them it is possible to explore all the possibilities concerning how the mechanism actually perform the function that must be explained. However, such explanations are not plausible in terms of the real components that perform the function, since they do not aim at describing these real component parts and operations.

Merely functional explanations are, therefore, not so good explanations for mechanists. But they can be used to set the early stages of the investigation of the mechanism; with time, they can be evaluated over and over so that it is possible to establish if they match with the empirical evidence and general information acquired through scientific research. In Craver's view, such models are merely heuristically useful for the construction and exploration of possible mechanisms, but they do not constitute adequate explanations (2007, p. 112). This means that models in this category are not sufficient to be qualified as good mechanistic models, since they do not offer information about the structural components performing the abstract functions, which is required in successful mechanistic explanations. Therefore, if a model of some cognitive capacity, for example, offers information solely about the abstract functional organization concerning that capacity, it cannot be considered a good mechanistic model, only a partial and bad one. In fact, one can argue that, since one of the central ideas of mechanistic explanations is to integrate structure and function, a mere functional explanation does not suffice for being considered mechanistic. Thus, functional explanations, without being complemented by structural explanations, are not mechanistic explanations at all.

The second category is *plausible mechanistic models*. Such models have a degree of plausibility since they match with the empirical evidence accumulated about the functioning of the natural mechanism under investigation, which can give support to the model. The third category is *actual mechanistic models*, and these models describe the functions being performed by the real mechanism in nature. The real relevant component parts of the mechanism and their operations are all well-known, as well as the causal interactions between them, i.e. the internal organization of the mechanism. Moreover, the external conditions affecting the performance of the natural mechanism are also well known.

The second assessment condition is related to the fact that a mechanistic explanation can be understood as a continuum that starts from an often very inaccurate model about how a given mechanism works to a complete detailed account of how this mechanism works. Here the progress in this continuum is a matter of accuracy in the information gathered about the real mechanism performing the particular functions, and there are also three important categories. The first can be called a *sketch* of the mechanism. The component parts, their operations and organization are still poorly understood and there are only general suppositions concerning what they might be and how exactly they might work. The possible parts might even not exist at all or not perform at all the operations that are being attributed to them. The model is also full of terms that are there just waiting for a more accurate explanation (these are called filler terms: to activate, produce, process, cause, encode, etc.), i.e. the terms indicate that some particular activity is being performed, but what activity exactly remains to be understood.

The second category is a *schema* of a given mechanism. This schema can be poorly or highly accurate, i.e. it has some degree of detail concerning the information about the actual mechanism performing the function. But what characterizes the schema is that researchers already have some more accurate (but still partial) idea about how the mechanism works. This means that some of the component parts and their operations are already understood, as well as the interactions between them with some degree of detail.

The third category concerning mechanistic models on this second assessment condition is the ideal of a *complete model* about how the mechanism works. In this category, all the details of the real mechanism under consideration are known and present in the model, as well as the behavior of the model in different environmental conditions. Weiskopf (2011, p. 317) points out that here the point is best seen as being about "grain and correctness". The ideal accuracy in terms of correctness involves the most correct information about all the parts of the system, and it includes no irrelevant components while it includes all the relevant components. The ideal knowledge in terms of grain involves the highest level of detail concerning all the sub-parts and sub-subparts, and so on, of the mechanism, i.e. information in terms of fine and coarse grain levels of understanding the whole system. Of course, this is a mere ideal, but it is or should be the ultimate aim of a mechanistic explanation. Once such a complete description is provided, we have all the relevant information concerning the phenomenon and the mechanism that performs it.

The third assessment condition concerns the distinction between: merely descriptive models, which can be useful for describing the components of a system and provide even some predictions about its behavior, while they are not explanatory; and explanatory models, which are more useful for control and manipulation and/or inform also how the system would perform under different or adverse conditions (Craver 2006). Thus, it is possible to say that "models become more explanatory the more they allow us to answer a range of counterfactual questions and the more they allow us to manipulate a system's behavior (in principle at least)" (Weiskopf 2011, p. 318).

The fourth assessment condition concerns the importance of background knowledge and local knowledge when investigating a given mechanism. It is necessary that the model of the biological mechanism under investigation does not contradict knowledge well-established in the field of biology, or that it does not contradict knowledge well established in the particular field that is trying to understand that particular biological mechanism.

Thus, at least four conditions can be established in order to assess good mechanistic explanations, following Weiskopf (2011, p. 318): (1) concerning whether mechanistic models are "confirmed or supported by the empirical evidence" that correctly indicates functional, but also structural components and operations; (2) concerning whether mechanistic models are informationally "accurate" in terms of the level of correctness and detail with which information about the actual system is represented by the model; (3) concerning whether mechanistic models are "genuinely explanatory" and not merely descriptive or predictive; and (4) concerning whether mechanistic models are "consistent with and plausible in the light of our general background knowledge and our more local knowledge of the domains as a whole".

In possession of these conditions for the assessment of successful mechanistic explanations in the biological sciences, including cognitive science, we can discuss whether all explanations in cognitive science need to be mechanistic in order to be successful scientific explanations. I will discuss this particular issue in detail in Chap. 6, Sect. 6.1.2, B.

2.2 Final Remarks

According to MTCBS and MTSEBS, a scientific explanation in the life sciences (e.g. biology, neuroscience, cognitive neuroscience, and cognitive science) starts from a general idea or sketch about how a biological mechanism performs a particular function (or gives rise to a particular phenomenon). The procedure to better understand this function is to decompose the mechanistic system in its working parts and subparts, to find out through localization what the operations related to these working parts are, and finally to understand the causal relations that are present in the system, i.e. its specific organization. In order to do this in the human brain, for instance, the techniques of recording of electrical activity and neuroimaging, for example, can be helpful, to the extent that they show how the parts can be decomposed, what are the activities related to each component and subcomponent.

This general way of proceeding also shows, according to the mechanistic framework, that this kind of explanation is very different from the traditional DNTSE. It does not use any deduction or derivation, and it does not typically appeal to general laws of nature. In fact, generalizations must be done with caution, because even when mechanisms clearly have similar coarse grained functions, as for instance the heart's function of pumping blood to the body in different species of animals, the differences between the structures of these hearts need to be considered: the heart of a human is different from a heart of a cat, a heart of a cow, and a heart of a dog, in some aspects, such as size, shape, capacity of pumping blood, heart rate in terms of contractions of the heart per minute (bpm), and so on. To establish general laws of nature in such a context of explanation is misleading, in the view of neo-mechanists. In their view, successful genuine scientific explanations in the life sciences are and should be constructed in terms of complex biological mechanisms and the functions they perform.

References

Bechtel, W.: Decomposing the mind-brain: A long-term pursuit. Brain and Mind **3**, 229–242 (2002)

Bechtel, W.: Reducing psychology while maintaining its autonomy via mechanistic explanations. In: Schouten, M., Looren de Jong, H. (eds.), The Matter of the Mind: Philosophical Essays on Psychology, Neuroscience, and Reduction, pp. 172–198. Blackwell Publishing, Malden, MA (2007)

Bechtel, W.: Mental Mechanisms: Philosophical Perspectives on Cognitive Neurosciences. Routledge, New York (2008)

Bechtel, W.: Constructing a philosophy of science of cognitive science. Topics in Cogn. Sci. **1**(3), 548–569 (2009a)

Bechtel, W.: Looking down, around and up: Mechanistic explanations in psychology. Philos. Psychol. **22**(5), 543–564 (2009b)

Bechtel, W.: Mechanism, modularity, and situated cognition. In: Robbins, Philip, Aydede, Murat (eds.) The Cambridge Handbook of Situated Cognition, pp. 155–170. Cambridge University Press, Cambridge (2009c)

Bechtel, W.: Identity, reduction, and conserved mechanisms: Perspectives from circadian rhythm research. In: Gozzano, S., Hill, C.S. (eds.) New Perspectives on Type Identity: The Mental and the Physical, pp. 43–65. Cambridge University Press, Cambridge (2012)

Bechtel, W., Abrahamsen, A.: Explanation: A mechanist alternative. Stud. Hist. Philos. Biol. Biomed. Sci. **36**, 421–441 (2005)

Bechtel, W., Wright, C.D.: What is psychological explanation? In: Symons, John, Calvo, Paco (eds.) The Routledge Companion to Philosophy of Psychology, pp. 113–130. Routledge/Taylor and Francis Group, New York (2009)

Craver, C.F.: Interlevel experiments and multilevel mechanisms in the neuroscience of memory. Philos. Sci. **69**(S3), S93–S97 (2002)

Craver, C.F.: When mechanistic models explain. Synthese **153**, 355–376 (2006)

Craver, C.F.: Explaining the Brain: Mechanisms and the mosaic unity of neuroscience. Oxford University Press, Oxford (2007)

Craver, C.F., Kaiser, M.I.: Mechanisms and laws: Clarifying the debate. In: Chao, H., Chen, S., Millstein, R.L. (eds.) Mechanism and Causality in Biology and Economics, pp. 125–146. Springer, Dordrecht (2013)

Craver, C.F., Tabery, J.: Mechanisms in science. In: Zalta, E.N. (ed.) The Stanford Encyclopedia of Philosophy, Spring 2017 edn. https://plato.stanford.edu/archives/spr2017/entries/science-mechanisms/ (2015)

Glennan, S.: The New Mechanical Philosophy. Oxford University Press, Oxford (2017)

Glennan, S., Illari, P.: The Routledge Handbook of Mechanisms and Mechanical Philosophy. Routledge, London (2018)

Hempel, C.G.: Aspects of Scientific Explanation and Other Essays in the Philosophy of Science. Free Press, New York (1965)

Hempel, C.G., Oppenheim, P.: Studies in the logic of explanation. In: Hempel, C.G. (ed.) Aspects of Scientific Explanation and Other Essays in the Philosophy of Science, pp. 245–290. Free Press, New York (1965, Originally published in 1948)
Krickel, B.: The Mechanical World: The Metaphysical Commitments of the New Mechanistic Approach. Springer (2018). https://doi.org/10.1007/978-3-030-03629-4
Machamer, P., Darden, L., Craver, C.: Thinking about mechanisms. Philos. Sci. **67**, 1–25 (2000)
McLaughlin, B.: The rise and fall of British emergentism. In: Beckermann, A., Flohr, H., Kim, J. (eds.) Emergence or Reduction? Essays on the Prospects of Nonreductive Physicalism, pp. 49–93. Walter de Gruyter, Berlin (1992)
Nagel, E.: The Structure of Science. Problems in the Logic of Explanation. Harcourt, Brace and World, Inc., New York (1961)
Piccinini, G.: Computationalism. In: Margolis, E., Samuels, R., Stich, S. (eds.) The Oxford Handbook of Philosophy of Cognitive Science, pp. 222–249. Oxford University Press, New York (2012)
Samuels, R., Margolis, E., Stich, S.: Introduction: Philosophy and cognitive science. In: Margolis, E., Samuels, R., Stich, S. (eds.) The Oxford Handbook of Philosophy of Cognitive Science, pp. 3–18. Oxford University Press, New York (2012)
Stephan, A.: Emergence—a systematic view on its historical facets. In: Beckermann, A., Flohr, H., Kim, J. (eds.) Emergence or Reduction? Essays on the Prospects of Nonreductive Physicalism, pp. 25–48. Walter de Gruyter, Berlin (1992)
Thagard, P.: Hot Thought: Mechanisms and Applications of Emotional Cognition. MIT Press, Cambridge, MA (2006)
Uttal, W.R.: The New Phrenology: The Limits of Localizing Cognitive Processes in the Brain. MIT Press, Cambridge, MA (2001)
Uttal, W.R.: Mind and Brain: A Critical Appraisal of Cognitive Neuroscience. MIT Press, Cambridge, MA (2011)
Uttal, W.R.: Reliability in Cognitive Neurosciences: A Meta-Meta-Analysis. MIT Press, Cambridge, MA (2012)
Weiskopf, D.: Models and mechanisms in psychological explanation. Synthese **183**, 313–338 (2011)
Woodward, J.: Scientific explanation. In: Zalta, E.N. (ed) The Stanford Encyclopedia of Philosophy, Fall 2017 edn. https://plato.stanford.edu/archives/fall2017/entries/scientific-explanation/ (2014)
Wright, C., Bechtel, W.: Mechanisms and psychological explanation. In: Thagard, Paul (ed.) Philosophy of Psychology and Cognitive Science, pp. 31–79. North Holland Elsevier, Amsterdam (2007)

Chapter 3
The Mechanistic Theory of Human Cognition

3.1 Central Theoretical Components of MTHC

3.1.1 Introducing the Theory: A Brief Overview of the Context of Its Genesis

MTHC[1] has been developed based on a broad physicalist context that dominates a vast amount of work in contemporary cognitive science, philosophy of cognitive science and philosophy of mind. Furthermore, the theory attempts to combine central views and notions present in traditional cognitive science with major views and notions present in certain fields of neuroscience concerned with cognition and human cognition. In this overview, I will discuss briefly these background aspects which are central for the general formulation and development of MTHC.

In the second half of the twentieth century many important philosophers argued more clearly for a new version of 'physicalism'[2] as a philosophical thesis about the ultimate nature of reality. Physicalism is the philosophical doctrine defended by those who believe, roughly, that everything in nature or reality is made of or constituted by physical things—from physical sub-particles, to molecules, brains, planets, societies and ecosystems, including, of course, human cognitive capacities. Notorious analytic philosophers such as Willard Quine, David Lewis, David Armstrong, Donald Davidson, Jerry Fodor and many others are all physicalists of one stripe or another. Physicalism is, to a large extent, the 'world view' of many contemporary analytic philosophers and contemporary natural scientists all over the world. It is intended to be a very general and abstract doctrine that is true for the whole world. In other words, it is considered by its proponents to be a thesis about the ultimate nature of

[1] This framework has ontological and epistemological elements, which will be discussed together in this chapter. MTHC considered exclusively from an epistemological point of view can be called *Mechanistic Theory of Scientific Explanations in Cognitive Science* (MTSECS).

[2] I use the word 'physicalism' to describe this philosophical position, instead of 'materialism' or 'naturalism'. For a discussion cf. Stoljar (2010, 2015), and Papineau (2015).

D. A. Leite, *The Twenty-First Century Mechanistic Theory of Human Cognition*, Cognitive Systems Monographs 41, https://doi.org/10.1007/978-3-030-63680-7_3

the world that we have "considerable and perhaps overwhelming reason to believe" (cf. Stoljar 2010, p. 13). Physicalism became something like a consensus position within analytic philosophy in the 1960s and has remained so ever since (Papineau 2001; Stoljar 2010).

Physicalism is also strongly related to an overwhelming acceptance of the method of the natural sciences as the only genuine and true general methodology of inquiry for the real phenomena in the world—thus the ontological thesis is often connected to an epistemological and a methodological one. According to Churchland (1986, pp. 2, 3), philosophy should be on a continuum with the empirical sciences.[3] The higher complexity of certain questions is regarded as a matter of degree and not of a qualitative difference. Theories are considered to be interesting just to the extent that they can be related to empirical observational facts. In this view, one finds also a rejection of so-called armchair philosophy and pure linguistic analysis as a method of philosophy; instead, it is argued, philosophy should stay close to the empirical sciences—even though there are divisions of labour, such divisions need not to be so strict, nor should they imply a radical difference in methodology. These ideas form the background for many influential proponents of MTHC (cf. this chapter, Sect. 3.1.3).

In addition to this broader philosophical development in the twentieth century, there were also particular developments of important new scientific theories concerning human brain and cognition/mind.[4] Firstly, there was the rise of cognitive psychology approximately in the end of the decade of 1950 and beginning of the decade of 1960, and cognitive science in the decade of 1970 (cf. Gardner 1985; Harnish 2002; Miller 2003; Sturm and Gundlach 2013; Thagard 2005, 2018). Secondly, there was the unification of the brain sciences (e.g. neuroanatomy and neurophysiology) under the term 'neuroscience'—also around the decade of 1960 (cf. Bechtel 2008, p. ix)—and its growth as a discipline until present times. For approximately two decades and a half (1960–1985), though, cognitive psychologists and neuroscientists remained largely isolated from each other (cf. Wright and Bechtel 2007, p. 43). The proponents of dominant computational approaches in cognitive science traditionally attempted to provide a general cognitive computational architecture, abstracting from its concrete implementation in physical devices, such as a brain. In other words, the computationalist strategy was "first to articulate a computational theory of cognition, and then to inquire into how the implicated computational processes might be carried out in the brain" (Cummins 2000, p. 133). However,

[3]The works of Quine, *World and Object* (1960), and Sellars, *Science, Perception and Reality* (1963), were very important in giving support for these views.

[4]Currently, it is far from clear what the distinctions between the concepts of 'mind' and 'cognition' are. The use of 'cognition' as synonymous of 'mind' is overwhelmingly present in the specialized literature (cf. e.g. Anderson 2015; Bermúdez 2014; Eysenck and Keane 2015; Frankish and Ramsey 2012; Gardner 1985; Goldstein 2015; Miller 2003; Neisser 1967/2014; Smith and Kosslyn 2014; Sternberg and Sternberg 2012; Thagard 2005, 2018; Von Eckardt 1993; Ward 2015). In spite of this acceptance, it remains controversial whether these two concepts can indeed be treated as having the same meaning. More systematic research in this respect is, therefore, needed. I will be using these terms as synonymous in the present work, only in order to make the discussions clearer.

approximately in the second half of the 1980s, this strategy started to change more dramatically. Already in 1986, Patricia Churchland calls for the integration of cognitive research and neural research in her book, *Neurophilosophy: Towards a unified science of the mind-brain.* She points out that:

> [...] top-down strategies (as characteristic of philosophy, cognitive psychology, and artificial intelligence research) and bottom-up strategies (as characteristic of the neurosciences) for solving the mysteries of mind-brain function should not be pursued in icy isolation from one another. What is envisaged instead is a rich interanimation between the two, which can be expected to provoke a fruitful co-evolution of theories and methods, where each informs, corrects, and inspires the other. (Churchland 1986, p. 3)

The aim of Churchland's book was to outline "a very general framework suited to the development of a unified theory of the mind-brain", as well as to encourage the interaction between "philosophy, psychology, and neuroscience, or more generally, of top-down and bottom-up research" (Churchland 1986, pp. 3–4).

In the same period, many scientists were already perceiving the scientific gap between brain research and cognitive research. Thus, a new discipline called "cognitive neuroscience" began to emerge with the aim of filling this gap (cf. Bechtel 2009d, p. 33). Around the 1980s the scientist Michael Gazzaniga established the *Cognitive Neuroscience Institute*, and in 1986 a book edited by Joseph LeDoux and William Hirst called *Mind and Brain: Dialogues in Cognitive Neuroscience* was published. In 1989, the *Journal of Cognitive Neuroscience* was founded by Gazzaniga and in 1994 he cofounded the *Cognitive Neuroscience Society*, "providing institutional identity to the new initiative" (Bechtel 2009d, p. 34). Besides, developments of new methods of brain research were also extremely significant for the growing of cognitive neuroscience. For instance, the development of methodologies providing images of the brain, such as positron emission tomography (PET) and functional magnetic resonance imaging (fMRI) in the 1990s. In this context, towards the final period of the century, on the one side, there was a great rising of cognitive neuroscience research (Baars and Gage 2013; Gazzaniga et al. 2014; Ward 2015) attempting to provide understanding on the biological and neural basis of cognition; and on the other side, there was an increasing turn of cognitive psychology and cognitive science towards neuroscience and biology (cf. Anderson 2015; Bermúdez 2014; Thargard 2018).

Pressures, tendencies and desires to increase this integration in present days are very strong. It is actually so strong that sometimes authors do not recognize anymore almost any difference between cognitive psychology and cognitive neuroscience. A clear example is found in the popular textbook *Cognitive Psychology: a student's handbook* (2015), written by Michael Eysenck and Mark Keane. In this book, the authors claim that a great change in recent years "has been the ever-increasing emphasis on studying the brain as well as behavior" and they use the term "cognitive psychology" to refer to this approach; they note, though, that the term "cognitive neuroscience" is increasingly used to describe the same approach (Eysenck and Keane 2015, p. xv). They claim, furthermore, that cognitive psychology as a science that aims "to understand human cognition" and its "internal processes", such as "attention", "perception", "memory", "language", "reasoning", by "observing the behavior" can include also more broadly "brain activity" as a "source of relevant information for understanding human cognition", since we are now convinced that "we need to study the brain as well as behavior while people engage in cognitive tasks"—after all "the internal processes involved in human cognition occur in the brain" (Eysenck and Keane 2015, p. 1). But since "cognitive neuroscience" just uses "information about behavior and the brain to understand human cognition", the authors conclude: "Thus, the distinction between cognitive neuroscience and cognitive psychology in the broader sense is blurred" (Eysenck and Keane 2015, p. 1).

In the equally popular textbook of John Anderson, *Cognitive Psychology and Its Implications* (2015), the field of cognitive psychology is defined as "the science of how the mind is organized to produce intelligent thought and how the mind is realized in the brain" (Anderson 2015, p. 1). This is in order especially because nowadays "we all know that the brain is the organ of the human mind" (Anderson 2015, p. 1). Anderson says that classic cognitive psychology (especially around the 1960s and 1970s) typically developed their theories and explanations of cognitive processes without "any reference to the brain" (2015, p. 9). However, he recognizes that there was a "shift" in the way cognitive psychologists approach cognition: they have "gradually broadened their approach [...] as they have begun to pay more attention to the nature of information processing in the brain" (Anderson 2015, p. 10). In Anderson's view, much theoretical development in cognitive psychology is traditionally "of the abstract information-processing sort" not because of a belief in some sort of "dualism", but because "until recently many believed that brain activity was too obscure to provide a basis for understanding human cognition" (Anderson 2015, p. 10). However, as he points out, "the steady development of knowledge about the brain and methods for studying brain activity" is slowly eliminating the "barriers to understanding the mind by studying the brain"; as a result, "brain processes are now being considered in almost all analyses of human cognition" (Anderson 2015, p. 10). The book has in fact a brief section called 'cognitive neuroscience', which is defined as "the study of how cognition is realized in the brain"; it has also many information concerning brain anatomy and activity, as well as discussions about neuroscientific methodology, which "now inform the study of human cognition, enabling us to see how cognition unfolds in the brain" (Anderson 2015, p. 10). In this way, understanding "the strengths and weaknesses of the human nervous system is a major goal in understanding the nature of human cognition" (Anderson 2015, p. 11). Interestingly, as we can note, his definitions of cognitive psychology and cognitive neuroscience are extremely similar (to say the least), making it very hard to find the difference between these fields and what exactly their scopes and limits are.[5]

Indeed, to integrate traditional cognitive science with traditional neuroscience trying to get the best of them in order to understand human cognition seems to many authors to be the next reasonable step to make. Perhaps, it is the only way

[5]Similar remarks concerning cognitive psychology and cognitive neuroscience are found in Goldstein (2015), Smith and Kosslyn (2014), and Sternberg and Sternberg (2012).

to completely understand the neuro-cognitive relationship. Currently, even particular computational cognitive architectures, such as the ACT-R of John Anderson (cf. Anderson 2007), or the SOAR of John Laird (Laird 2012), or the SIGMA of Paul Rosenbloom (Rosenbloom et al. 2016), or the SPAUN of Chris Eliasmith (cf. Eliasmith 2013), or the ALPHAGO developed by Google DeepMind through the framework of deep reinforcement learning (cf. Mnih et al. 2015), are developed within a considerable knowledge of neural activity, cognitive functions and how to artificially integrate at least some of them. Of course, these architectures are very different, but all of them are build taking into consideration that knowledge, even if it is used by each of them in different degrees and forms.

MTHC has been proposed in order to provide such integration in a more precise theoretical way and it has achieved great recognition and adherence. The new framework has been applied to human cognition in an attempt to provide a general theory of human cognition and of its relation with neural activity. Many authors have argued that the theory gives a significant contribution for the problem of relating human cognitive and neural phenomena and for the problem of integrating, at the same time, cognitive science and neuroscience into a large scientific enterprise (cf. Bechtel 2008, 2009a, b, c, d, 2010; Boone and Piccinini 2015; Craver 2007; Piccinini and Craver 2011; Thagard 2006, 2009, 2019; Zednik 2018).

3.1.2 Human Neuro-Cognitive Mechanisms

Explaining how human cognitive structure and function are integrated is a hard task. Classical cognitive science aimed ultimately at such integration, but its focus was generally on functions, traditionally neglecting the structure, i.e. the physical implementation. Cognitive scientists have generally focused on discussing issues such as whether representations are symbolic, syntactical and related structurally according to rules, and whether there is any kind of 'language of thought'. Because of this, they were often criticized by authors who believe that investigations of neural activity need to play a major role in the field. Neuroscientists, on the other hand, have generally focused on the activity inside the nervous system and adopted largely the map-like strategy for characterizing operations in the brain that 'represent' external stimuli, i.e. what physical and chemical processes (such as the spiking of neurons) take place in the brain when a certain kind of stimulus is presented. Authors following this approach were also criticized for neglecting more complex cognitive activity. Given this divergence concerning scientific strategies of investigation about human cognition, there has been traditionally a gap between cognitive and neuroscientific studies and interests. The mechanistic theory addresses this difficulty and tries to offer a solution for it. It attempts, ultimately, to generally integrate these two major forms of investigating human cognition.

Many neo-mechanistic accounts concerned with applications to aspects of cognitive science and human cognition have been proposed in the specialized literature. One of the most influential and detailed proposals is developed by William Bechtel, who also tried to apply it to specific cognitive phenomena, such as memory consolidation and visual perception. This is the reason why the discussion concerning human neuro-cognitive mechanisms in this sub-section of the present chapter is mostly based on his work. In the following sub-sections of this chapter, the discussion of the main theoretical elements of the mechanistic theory of human cognition has Bechtel's work as a primary source. However, the work of Carl Craver is used to a large extent, since he also provides important and detailed contributions to the issue of human cognition from a neo-mechanistic point of view. The work of influential neo-mechanists, such as Gualtiero Piccinini, Paul Thagard, and others, are used in the following discussions, as well. The integration of these different accounts will provide the most crucial elements to construct a formulation of MTHC as complete and accurate as possible. This formulation can be then critically analysed and evaluated without distortions and mischaracterizations.

In this sub-section, therefore, my aim is to present and describe in general lines what a human neuro-cognitive mechanism essentially is. I will use Bechtel's work merely to illustrate the main characteristics of the neo-mechanistic account of a human neuro-cognitive mechanism and its main elements.[6]

Bechtel's version of MTHC considers the (human) "mind/brain" as a "set of mechanisms for controlling behavior" (2008, p. 159). He explains cognitive/mental phenomena (e.g. perception, attention, memory, problem solving, language) particularly by mechanisms characterized as "information-processing mechanisms"; thus, to put it clearly: cognitive/mental mechanisms are information-processing mechanisms (Bechtel 2008, p. xi).

In contrast to what happens most often in biology, in which both the phenomena itself and the operations that explain them are characterized in terms of physical transformations of material substances, mental mechanisms are characterized according to two different forms (or perspectives):

> The performance of a mental activity also involves material changes, notably changes in sodium and potassium concentrations inside and outside neurons, but the characterization of them as mental activities does not focus on these material changes. Rather, it focuses on such questions as how the organism appropriately relates its behavior to features of its distal environment – how it perceives objects and events distal to it, remembers them, and plans actions in light of them. The focus is not on the material changes within the mechanism, but rather on identifying more abstractly those functional parts and operations that are organized such that the mechanism can interact appropriately in its environment. Thus, mental mechanisms are ones that can be investigated taking a physical stance (examining neural structures and their operations) but also, distinctively and crucially, taking an information-processing stance. (Bechtel 2008, p. 23)

This basically means that mental phenomena can be described in terms of the physical structure with the physical-neural operations responsible for performing them, or in terms of 'information processing', i.e. in terms of abstract functions, which relate the phenomenon to the environment, often using the 'input, cognitive process, output' notion, where the cognitive process realizes the function of connecting the input

---

[6]A more detailed discussion and a comparison with other influential views advocated by other neo-mechanists, however, is carried out in Chap. 6, Sect. 6.1.3.

(environmental stimuli) with the output (behaviors). However, these two different forms of characterizing cognitive processes refer to the same phenomena, i.e. these are two different ways of talking about the same thing.

Bechtel accepts the idea of characterizing 'information' in terms of "regular effects of a cause that make it possible to infer features of the cause from features of the effect" (2008, p. 24). Information, according to this account, is thus a causal notion. The effect carries information about the cause, and this is the reason the effect is important: not because of its intrinsic properties, but rather because it contains information about its cause. Therefore, the mechanistic theory is committed to cognitive information processing, i.e. to the overwhelmingly held notion that cognition is a kind of information processing, which has been central in cognitive science.

The notion of cognitive (mental) representation is also one of the most central in MTHC. Bechtel argues that representations should not be abandoned in cognitive science and related sciences dealing with the 'mind/brain'; the fact that "cognitive theories invoke representations" is what distinguishes "the cognitive tradition from its behaviorist predecessor in psychology" (2008, p. 159). And the author considers the notion as useful and fruitful in cognitive science. However, the notion of 'cognitive representation' this position relies on is particular. 'Cognitive representation' here is understood in terms of 'effects representing their causes'. According to this view, in simpler nervous systems representations can be considered as a coordination of motor activities when some external stimuli (i.e. information) is presented to the senses; in more complex nervous systems representations can be used to coordinate more complex functions (Bechtel 2008, p. 159). There are two major features of representations which are necessary in the account of information processing mechanisms: the 'vehicle', i.e. the thing that represents something; and the 'content', what is being represented, i.e. that something which the vehicle "carries information about"—for instance, the "sound wave reaching a person's ear after lightning is the vehicle, and the information carried about the electrical discharge is the content" (Bechtel 2008, pp. 24–25). Other examples of representational vehicles are spoken or written words, diagrams, maps, and pictures. They 'designate something else' and all of them can be described according to their physical features. The content in these cases are what the words, maps, diagrams and pictures refer to.

Accordingly, mental/cognitive mechanisms are characterized as information processing mechanisms because "the operations within them produce and change informational vehicles in ways that correspond to the informational content they carry about things and events external to themselves" (Bechtel 2008, p. ix). It is necessary, Bechtel argues, to consider both the operations occurring inside the mental mechanism and "how what is operated on serves to stand in for things and events in the external world with which an organism is coordinating its behavior" (2008, p. xi). In this sense of representations, thus, it can be said that specific regions in the brain, with their neurons, connections and activities, represent the muscles that they control; as well as external particular stimuli from the environment are represented in brain areas if neurons in these areas fire when the stimuli are presented. What 'transmits information' here is the nerve fiber and 'information is processed' when neurons communicate with each other through electrical and chemical activities.

Physical information about the environment is encoded and decoded in a physiochemical way and informational process is physiochemical process. Some examples of information processing are thus the following operations of the brain: "ganglion cells in the frog's retina that serves as bug detectors"; "cells in primary visual cortex that [function] as edge detectors"; and "the capacity of Purkinje cells in the cerebellar cortex to learn motor patterns" (Bechtel 2008, p. 25).

Accordingly, in Bechtel's view, the best strategy for linking the representational/information processing perspective to the physical/implementational perspective is "to maintain that operations in the ultimate algorithmic decomposition correspond to specifiable physical processes characterized in the implementational account" (2008, p. 28). Regarding this aspect, computers, according to him, and the relationship between hardware and software provide the best example of "how to bridge from the representational/algorithmic perspective to the implementational perspective": what is necessary is just "a proper match between operations specified in the machine code into which a program is compiled and the physical operations in the computer itself" (Bechtel 2008, p. 31). In this way, "one could develop software to perform tasks and know that the computer once programmed would behave appropriately" (Bechtel 2008, p. 31). As he puts it:

> [...] the implementation perspective is a different perspective on the same entities and processes described from the information processing perspective. In the language I used above, the implementational perspective focuses on the vehicle, not its content, whereas the representational and algorithmic analysis focuses on the content, not the vehicle. [...] One could, in principle, equally track the processes in the mechanism in terms of the vehicles or the contents. However, insofar as the objective is to explain how the mechanism enables an agent (organism) to coordinate behavior with its environment, the characterization of processes in terms of content is usually the more informative. (Bechtel 2008, p. 34)

As we can see, the framework incorporates notions such as 'algorithm', 'computational processing' and 'software' to refer and characterize aspects of human cognition. However, extremely little is actually said in support of this commitment to 'cognitive computation' and no detailed characterization is offered in Bechtel's work. One can thus note that there is a commitment to cognitive computation; however, this commitment is not so clearly articulated in the theory. Similarly, Craver (2005, p. 390) points out that "to integrate computation in the hippocampus with lower levels, one would show how the networks of cells are organized such that they execute the relevant computation in the hippocampus". This remark certainly does not tell much. However, one can clearly see that computation depends on the activities of a network composed by many cells in particular regions of the brain.

Therefore, it is clear that the neo-mechanists prefer a characterization of cognitive representations, information processing and computation more aligned with the neuroscientific work on the topic. As Bechtel sees it, the work in neuroscience concerning these notions presents a less problematic relation between vehicle and content.[7] He points out that "the neuroscientific strategy for identifying representations in the brain is relatively straightforward", since they "seek to determine what sensory stimulus or motor response is causally linked with the activity of particular neurons"; as a result, "neuroscience representations are more clearly grounded in the causal nexus relating organisms to their environments than are those advanced in cognitive science" (Bechtel 2008, p. 186).[8]

Bechtel's views on human cognitive/mental mechanisms are also complemented by the frameworks of *dynamical systems theory* and *control systems theory*. In his view, traditional forms of understanding the relationship between information processing and representations in cognitive science, such as "classical information processing and neural network models", are wrong, since they limit themselves to syntactical aspects of cognitive representations, neglecting issues regarding content (Bechtel 2009a, p. 556). He argues that "a profitable framework for thinking about the representations results from viewing mental mechanisms as control systems" (Bechtel 2008, p. xi). In his view, information processing is not enough to account for the content of representations. For this, control theory is necessary (Bechtel 2009a, p. 564). In this way, for the author, vehicles and contents can be properly related. This means that "insofar as cognitive mechanisms are information-processing mechanisms, cognitive science needs an account of how the representations invoked in cognitive mechanisms carry information about contents" (Bechtel 2009a, p. 548).

The author's approach considers cognitive information processing mechanisms as different from other biological mechanisms, which merely "operate by transforming internal states much as a cell transform glucose into carbon dioxide and water" (Bechtel 2009a, p. 557). Rather, since mental states have intrinsic content, the approach intends "to treat content as an essential aspect of representations (mental and otherwise) and give serious attention to the challenge of incorporating content into accounts of information processing" (Bechtel 2009a, p. 557). And, in his view, "control theory offers the needed perspective on the relation of representations to contents" (Bechtel 2009a, p. 548). That is, representations are features of control systems; they play a crucial role in systems that control other systems, since "they can achieve appropriate control only by representing relevant information" (Bechtel 2008, p. 161).

In the simplest forms of control systems, there is a closed loop control using feedback loop, which is considered a way of achieving control in mechanical and biological systems. What is being controlled is called the 'plant' and, "in negative feedback control, information about the effects of the operation of the plant is fed back so as to alter in appropriate ways the operation of the plant" (Bechtel 2008,

[7]Bechtel has indeed a strong background in biology and in work with neural networks. Thus, his preference for the notion of cognitive information processing, representations and computation more aligned with notions widespread in neuroscience should not be surprising at all.

[8]Bechtel belongs to an old tradition of thinking that considers 'the brain to be the seat of the soul', i.e. it is in terms of particular neural processes that one shall understand all particular human mental/cognitive activities. This tradition is represented, for example, by Julian Offray de La Mettrie, Franz Joseph Gall, Alexander Bain, David Ferrier, John Hughlings Jackson, Paul Broca and the 'localizationists' from the 1960s till present days (cf. Wright and Bechtel 2007).

p. 193). In the case of organisms, the brain is the responsible for the control—it is the controller. Information about a certain process is fed back through the system to regulate the behavior; for instance, information from the sensory system is fed back to the motor system in order to regulate its activity. In this sense:

> Representations will be found wherever control systems secure information about external phenomena, but this is entirely appropriate. In order to understand such systems, we need to be able to identify how information about a distal object or event is made available for the control system to use. The vocabulary of representation, of re-presenting what is not immediately present, accurately captures what is happening. Moreover, appeal to representations is not simply a means for us to interpret such systems. The systems themselves [...] would not work if information about the distal circumstances were not being made available within the control system. So representations are a central feature of the systems themselves. (Bechtel 2008, pp. 194–195)

However, Bechtel himself recognizes that while this approach works well for phenomena studied by neuroscience it is still more problematic for what is normally taken as typical cognitive phenomena "such as problems solving and planning" (2008, p. xi).

Bechtel (2016, p. 1287) also explores the idea of "neural representations" in neural mechanisms that process information, claiming that "characterizations of neural activity as representational contributes to the development of mechanistic accounts, guiding the investigations neuroscientists pursue as they work from an initial proposal to a more detailed understanding of a mechanism." Bechtel also emphasizes that: "Construals of neural activity as representations are not mere glosses but are characterizations to which neuroscientists are committed in the development of their explanatory accounts" (2016, p. 1287). In his view, neuroscientists want to determine what content neural representations have and how they represent this content. Contemporary neuroscience, thus, often integrates vehicle and content. This cognitive representational content, in turn, helps research in order to determine neural representational vehicles, but both are subject to scientific revision. Attributing cognitive content to neural representations is, thus, "a first step in articulating an account of mechanism for processing information" (Bechtel 2016, p. 1291). In fact, Bechtel states, "claims about representations can be viewed as instances of [type mind-brain] identity claims since what researchers are doing is identifying constituents of [neural] mechanisms as [cognitive] representations" (2016, p. 1291; cf. this chapter, Sect. 3.1.3). The example offered here refers to when a given population of neurons represent information in a "particular manner" related to some "source" (Bechtel 2016, p. 1291).

In this paper, Bechtel also emphasizes the role of control systems theory applied to neuroscience. He claims that some mechanisms in the brain are involved in regulating or controlling other parts of the brain or organs within the organism; in this way, the organism coordinates its behavior with features of the environment. The author states that:

> The brain is functioning as a control system and control processes require information about the plant and environment (including on occasion what conditions were like in the recent to distant past) that the controller can use in developing plans for action (that may not be

> executed until some time in the future). An important part of what neuroscientists want to understand about the brain is how it contributes to controlling processes within the organism or its behavior in the external world. It is in this context that identifying representations and their content becomes critical—it is as they represent entities and processes external to the brain that mechanisms within the brain figure in these control processes. (Bechtel 2016, p. 1292)

To illustrate this, the author uses the example of information-processing cognitive mechanisms that rodents employ in navigating their environment, together with the role played by place cells in the hippocampus. These place cells generate action potentials primarily when the organism is in a particular region of its local environment. The action potentials of these neurons are thus interpreted as representing that particular location.

Bechtel's approach is influenced by a variety of theories and frameworks, as he himself acknowledges (2009a). Based on this, he makes some useful contributions in order to clarify what exactly human neuro-cognitive mechanisms are and how exactly they relate to classic constructs in cognitive science, such as cognitive representations, information processing, and computation. However, these contributions, as one can note, are typically very general suggestions about how to fulfill the 'neuro-cognitive explanatory gap'. In sum, a more elaborated framework that shows how to apply such approach especially to higher-level complex cognitive activities is still missing.[9]

3.1.3 Neuro-Cognitive Identity Relations

In the background of MTHC lays the attempt to solve the difficult problems concerning human cognition within a general physicalist framework, i.e. roughly the view that everything in reality is constituted by physical things (cf. this chapter, Sect. 3.1.1). At the heart of this attempt we find the commitment of the theory to a version of the classic type-type identity theory between cognitive and neural processes. In a more clear and systematic way, the idea of establishing identity relations between cognitive and neural phenomena was spelled out through a traditional well-known program in philosophy of mind developed at the end of the 1950s, namely the classic mind-brain type identity theory, whose most prominent advocates are Place (1956), Feigl (1958), and Smart (1959). The identity theory became unpopular during the 1960s and 1970s, but given the advancements and success of neuroscience and cognitive neuroscience, and the turn of cognitive psychology and cognitive science towards brain research in the end of the last century, among other reasons, the identity thesis became once again a major option, but with modifications (cf. Gozzano and Hill 2012; Thagard 2018).

This commitment involves also three other important theoretical elements: (1) the idea that these neural processes and neural mechanisms that perform them can be decomposed in sub-parts, and these subparts further decomposed in sub-sub-parts,

[9] I will make a more detailed discussion of this point in Chap. 6, Sect. 6.1.3, C.

i.e. a mechanistic compositional relation (cf. this chapter, Sect. 3.1.4); (2) the idea that there are multiple hierarchical levels of decomposition in a mechanism (cf. this chapter, Sect. 3.1.5); (3) the idea that relevant autonomous processes of causation happen in all these different levels (cf. this chapter, Sect. 3.1.6). What follows is a discussion in detail of each of these theoretical components of MTHC.

Bechtel claims that the scientific disciplines which aim to explain cognitive activities recognize that "in some way these activities depend upon our brain" (2008, p. ix). Or to put in another way: "Psychological phenomena are realized in brains comprised of neurons" (Bechtel and Wright 2009, p. 118). This means that cognitive phenomena need to be explained somehow in a physical (neural) way. Bechtel indeed describes himself as adopting the "perspective" of "naturalism" (2008, p. x), or belonging to a group of philosophers of science that defend a "naturalist approach"; this "naturalist tradition" has its roots in the work of Willard Quine (1908–2000—who is considered to be one of the fathers of contemporary physicalism, cf. Stoljar 2010) and investigates the particular domains of science "in the manner in which scientists investigate phenomena in their own domains of inquiry" (2009a, pp. 550, 551).

However, Bechtel does not explicitly claim that he is a physicalist, or that he uses 'naturalism' as synonymous of 'physicalism', as many do. Nevertheless, Craver and Tabery (2015, Sect. 2.5) put the commitment in the clearest way: "[…] many mechanists opt for some form of explanatory anti-reductionism, emphasizing the importance of multilevel and upward-looking explanations, without rejecting the central ideas that motivate a broadly physicalist world-picture". There is, therefore, a kind of ontological monist physicalism that underlie the entire neo-mechanistic theory of human cognition, while at the same time the most significant proponents of the framework attempt to defend an anti-reductionist epistemological (explanatory) neuro-cognitive pluralism, i.e. roughly, the idea that psychological/cognitive science explanations do not completely reduce to explanations in the neurosciences.

In this physicalist ontological context, it is also clear that mechanistic explanations of human cognition aim to account for cognitive phenomena in terms of neural systems, their physical components and physical operations organized appropriately. Given this, there is no space for the kind of emergentism that introduces something "spooky" and "radically new" into the universe, as well as for "spooky metaphysical posits such as vital forces", in the framework (Bechtel 2008, p. 129; cf. Craver 2007, p. 16). Consider as well the following remarks made by some neo-mechanists concerning the nature of human cognition:

> There is no evidence that souls or entelechies exist. They cannot be detected by measuring devices, let alone with multiple methods embodying different theoretical perspectives. There are no clear criteria for determining when souls and entelechies are present or absent, and there are no clear criteria for individuating souls and entelechies (that is, clear and objective criteria according to which one could count them). We cannot intervene with predictable outcomes to change souls and entelechies, and we cannot use them to intervene in other states of affairs. For all these reasons, we are justifiably suspicious of claims that such things exist. But none of these reasonable criteria fails for higher-level items in neuroscience. Molecules, neurons, brain regions, and brain systems all clearly satisfy these standards. (Craver 2007, p. 15)

> The brain is composed of neurons. Neurons transmit signals in the form of action potentials. They communicate across electrical and chemical synapses. They are composed of a complicated array of cytoplasmic molecules. They have characteristic ways of generating electricity, of repairing themselves, and of eliminating waste products. They are organized into networks of cells that make up systems, many of which have widely conserved patterns of organization. These diverse items constitute the ontic store of contemporary neuroscience: the set of stock-in-trade items out of which models of mechanisms can be built. (Craver 2007, pp. 249–250)

> We do endorse reductionism in the sense that every concrete thing is made out of physical components and the organized activities of a system's components explain the activities of the whole. Setting aside dualism and spooky versions of emergentism, we take these theses to be uncontroversial. (Piccinini and Craver 2011, p. 284)

> […] cognition is explained (at some level) by neural network activity. But this is a truism – or at least it should be. The brain is the organ of cognition, the cells that perform cognitive functions are (mostly) neurons, and neurons perform their cognitive labor by organizing themselves in networks. (Piccinini 2012, p. 241)

These passages clearly indicate that they endorse some sort of ontological monist physicalist position concerning human cognitive capacities. As they see it, there are no "gods and goblins" (Craver 2007, p. 15) in this approach; there is no space for ontological dualism (be it of substances or properties, fundamental or emergent), pluralism, or non-physicalism (e.g. idealism or neutral monism) of any sort.

Craver states the same point in yet another way. In his view, real parts (i.e. not fictional) "have a stable cluster of properties, they are robust, they can be used for intervention, and they are physiologically plausible" (Craver 2007, p. 131). The criteria of stable cluster of properties refer to the capacity of parts to operate regular biochemical and physical processes; the criteria of robustness refer to the possibility of the parts, operations and processes to be detectable by means of causally and theoretically independent physical devices and techniques; the criteria of intervention means that one should be able to manipulate directly a component operation in order to change another; finally, the criteria of physiological plausibility means that mechanisms, their parts, and their activities should exist under normal conditions in the physical spatial-temporal world (Craver 2007, pp. 131–132). Moreover, as anything physical within the space-time, mechanisms have spatial and temporal organization. The former involves for instance locations, sizes and shapes; the latter, order, frequency and duration (Craver 2007, pp. 137, 138). Therefore, the commitment to ontological physicalism and to the idea that human cognitive phenomena are ultimately neural phenomena are clear in MTHC.

The important question is, however, how exactly these commitments are sustained. One of the clearest ways of assuming an ontological monist physicalist position in the mind-brain debate is to get rid of mind-brain correlations through the attempt to establish a relation of identity between these two phenomena. This is precisely what MTHC does, and it is one of the most important aspects of the theory. The mechanistic theory, in its clearest form, advances basically an attempt to overcome problems with traditional arguments and put the mind-brain type identity thesis in more solid grounds. Bechtel and McCauley (1999; cf. Bechtel 2002, p. 236ff.) argue for what they call the heuristic identity theory (HIT), a variant on the traditional psycho-neural type identity theory. In this account, there is an identity between the types of cognitive and neural mechanistic processes. As Bechtel states: "Underlying attempts to localize cognitive operations in brain structures is the assumption that there is an identity relation between particular mental mechanisms and neural mechanisms" (2008, p. 70). The thesis of the neuro-cognitive type-type identity means that 'some particular type of function performed by a given type of neural mechanism is identical to a particular type of cognitive function'. For instance, a type of neural mechanism *W* performs a particular neural function N_1. This function N_1 is identical to a particular cognitive function C_1. The cognitive function related to a kind of pain, for example, on that account, is considered to refer to a kind of process performed by a particular mechanism in the brain (probably some type of somatosensory activity performed by a given neural network).

Another important aspect is that the statements refer to 'heuristic identities'. This means that the statements of identity here are not logical or mathematical necessary ones, such as $2 + 2 = 4$, or $x = x$. They are rather empirical and contingent. Identity claims in science, the neo-mechanists argue, play typically a heuristic role in the sense that they are adopted as hypothesis to guide further investigations, like in the following cases: H_2O is identical to 'water', 'molecular kinetic energy' is identical to 'heat', and 'electromagnetic radiation' is identical to 'light'. Identities, on this view, are not conclusions of scientific work. They are rather used as hypothetical premises that help scientific research to advance: they offer hypotheses that could be further investigated in order to provide empirical evidence for the identity.

In this sense, heuristic identities can also be improved with scientific development. Bechtel explains that heuristic identity claims are particularly useful for relating function and structure in mechanisms. When identities are proposed there can be structures which have no functional correlates, or functional relations with no structural correlates. Consequently, the research will aim to determine if the corresponding component parts and operations can be found. Thus, heuristic identities have the virtue of guiding "not only the elaboration of the two perspectives which are linked by the identity claim, but it can use each to revise the other" (Bechtel 2008, p. 71). This means that in mechanistic explanations it is not necessary to identify the component operations correctly before they attempt to localize them in the brain: "As long as the initial hypothesis as to the operation performed is even roughly in the right ballpark, an identity claim can play a fruitful role in generating evidence that leads to revisions and refinements of the initial claim" (Bechtel 2008, p. 71). McCauley (2012) also agrees on this point. In his view, identities that span different levels of explanation are indicated by scientists in order to provide bridges for investigations at these different levels of analysis—in this way, theories, experimental strategies and evidence can be imported from one level to another (McCauley 2012, p. 192).

Interestingly, Craver rejects the idea of using identity relations to characterize his mechanistic account: "Integrating mechanistic levels is not a matter of establishing identities across levels" (2007, p. 258). There is, therefore, a major difference between the formulations of the theory according to these two different influential authors.[10]

3.1.4 *Neuro-Cognitive Compositional Relations*

Another central aspect of MTHC is its theoretical commitment to 'compositional relations' (cf. Chap. 2, Sect. 2.1.5) in order to describe human cognitive biological mechanisms. The idea endorsed here is that all human cognitive mechanisms have at least two levels, the higher-level of the whole mechanism and the lower level of the components (single components and sets of components smaller than the entire given whole mechanism). That basically means that all mechanisms of this kind are necessarily decomposable in a certain number of parts. What bridges the mechanistic levels and makes the 'interaction' possible is the "compositional relations between parts and wholes in a mechanism" (Bechtel 2008, p. 146). This mechanistic compositional relation is a kind of "part-whole relation" (Craver 2007, pp. 6, 8) and it is explanatory, since the functions of the mechanism are explained in terms of its components and their organization. In this sense, the working parts, operations and a particular organization of a given mechanism compose this mechanism.

This notion of 'composition' is used by neo-mechanists as synonymous of the notion of 'constitution'. In the same way, thus, one can say that the component parts of a given mechanism constitute this mechanism. As the mechanists point out, the notion of compositionality is not meant to be used in 'etiological' mechanistic causal explanations, i.e. the explanation of an event by its antecedent causes; it is rather 'constitutive', i.e. the parts and their organization constitute the whole and thereby explain the phenomenon/behavior under consideration (Craver 2007, pp. 8, 108, 128). Craver also claims this mechanistic constitutive relation is "symmetrical" since alteration in the phenomenon produce alterations in the component parts and alterations in component parts produce alteration is the phenomenon. In his view, thus, "all constitutive dependency relationships are bidirectional" (Craver 2007, p. 153).

3.1.5 *Neuro-Cognitive Hierarchical Mechanistic Levels*

What follows from this compositional view is that biological mechanisms in neuroscience and cognitive science can also be considered to be multilevel (i.e. they can be composed by structures which belong to different levels) and hierarchically organized (i.e. these structures belong to lower or higher-levels) (Bechtel 2008, pp. 21, 22; Craver 2007, pp. 2, 9, 163).

[10] I make a critical discussion of identity relations in the mechanistic framework for cognitive science in Chap. 6, Sect. 6.1.3, B).

Indeed, it is common to see authors stating that in nature there are many different levels of organization. For instance, there are the levels of subatomic particles, atoms, molecules, chemical substances, cells, simple organisms, complex organisms (with complex organs like brains), complex societies of organisms (like communities of ants, families of chimpanzees, and human societies), complex ecosystems, complex solar systems, and so on. When considering the brain structure, it is also common to think in terms of levels, e.g. levels of molecules, neural cells, simple neural networks, larger neural network systems, etc. Similarly, when considering psychological phenomena, notions of levels are often used as well. For instance, when one talks about lower-level cognitive functions or capacities and higher-level cognitive functions or capacities. Or when authors argue that psychological explanations in terms of computations over representations are in a higher-level of analysis than the implementation level of the physical system that is responsible for performing such computations. It is thus clear that a notion of levels is important not just for science in general, but also for cognitive science and neuroscience. These levels reflect the complex reality of the physical world and of many phenomena occurring in it. The difficulty is that the notion of levels is highly complex and controversial.

This notion of levels in science is a complicate matter since the term ‘level’ is very ambiguous. More systematic discussion about levels in science lead back at least to the classic paper of Oppenheim and Putnam (1958) where the authors argue in favor of the unity of science as a mean to counterbalance scientific specialization and achieve integration of scientific theories, encouraging thus a unified body of knowledge. In this paper, they try to relate explanations of human behavior to explanations of neural activity; explanations concerning the behavior of individual cells (including neurons) to biochemical explanations; explanations of molecular activity (including macromolecules that constitute living cells) to atomic physics (Oppenheim and Putnam 1958, p. 7). The levels correspond to the scientific disciplines that inquire about them. Physics is the most basic science at the bottom level. In higher-levels there are chemistry, biology, neuroscience, psychology, and sociology. In this account, there are multiple levels of organization, but scientific fields should be unified having physics as its base.

However, this account of levels and the scientific model of explanation behind it have been extensively criticized. For instance, it is often said that physics deals with atoms and subatomic particles, but it deals as well with planets, solar systems and galaxies. Are solar systems and subatomic particles then in the same level? It seems not. The same applies to biology which inquiries about a variety of phenomena, such as the components of genes and highly complex ecosystems. Therefore, understanding of levels in terms of scientific fields appears to be extremely problematic. Many accounts of levels in science were proposed after Oppenheim and Putnam’s account, and it is very common to see in the literature authors writing about levels of analysis, organization, size, explanation, description, complexity, realization, etc. Unfortunately, the current accounts of levels available still face sets of difficult problems. Notwithstanding the account of levels in mechanistic explanations still present some difficulties, it is arguably one of the most plausible in present days, in the view of neo-mechanists. Moreover, to give a systematic account of levels is at the same time indispensable, especially for a mechanistic integrative proposal of integrating neural and cognitive processes.

According to the neo-mechanistic account of levels, when one looks to the actual work of developing explanations done by cognitive scientists and neuroscientists, one observes that these explanations refer to a large variety of phenomena, e.g. the behaviors of organisms, the processing of functions of brain systems, the representational and computational properties of brain regions, the electrophysiological properties of nerve cells, and the structures and conformation changes of molecules (Craver 2007, p. 9). It is easy to note then that the explanation "oscillates up and down in a hierarchy of mechanisms to focus on just the items that are relevant to different phenomena or different aspects of the same phenomenon" (Craver 2007, p. 10). Thus, "there is no single neural level, or neurophysiological level, or neuroscientific level of explanation. Neuroscientific phenomena span a hierarchy of levels" (Craver 2007, pp. 10–11). At the different levels of organization, there are different causal processes occurring in mechanisms. This is why the levels in mechanisms also reflect the multitude of scientific areas that are supposed to explain them. Thus, we have the commitment to epistemological (explanatory) pluralism (anti-reductionism). In the case of cognitive biological mechanisms, for instance, we have frequently, among others, molecular neuroscience, cellular neuroscience, functional neuroanatomy, computational neuroscience, cognitive neuroscience, and cognitive psychology. However, one cannot know in advance how many levels will be necessary to explain a particular mechanism, since it depends on what the phenomenon being considered is.

At any rate, in mechanistic explanations it is "the set of working parts that are organized and whose operations are coordinated to realize the phenomenon of interest that constitute a level" (Bechtel 2008, p. 146). One starts the investigation at the highest level, i.e. the level of the whole mechanism and the phenomenon it produces. The working parts of this whole mechanism must be investigated at a lower level. If a working part is decomposed in further working parts, then each of these parts must be studied at a lower level, and so on. The fundamental level is reached at the last level that is still relevant for explaining the works of the mechanism. In this respect, Machamer et al. (2000, p. 13) remark that the fields of molecular biology and neurobiology do not normally regress to the quantum level to talk about the activities of chemical bonding, for instance. It would be thus rare if a scientist working at this level would be driven to lower levels; this could just happen in some special case, due to some special anomaly; but usually this does not happen. Of course, again, it depends on what the phenomenon being explained is. Thus, the theory is very flexible in this respect. Consequently, the question 'how many levels there are?' cannot have an absolute answer. It depends on how much levels are necessary to give a complete relevant explanation of the mechanism and the functions it performs.

3.1.6 Neuro-Cognitive Pluralist Causation

A central idea in MTHC, related to the multilevel aspect of human cognitive biological mechanisms, is to propose a kind of general explanatory pluralism which recognizes the importance of the analysis of multiple levels of organization in nature and provides a middle ground between those that want to reduce all the relevant levels to a basic and primary one and those that want to avoid this reduction creating a level where there are fictional metaphysical entities and activities (cf. McCauley and Bechtel 2001, p. 736).

This notion of multilevel biological mechanisms directly involves a particular notion of causal pluralism. According to the proponents of the view, causation across multiple levels happen through 'mechanistically mediated effects' (Bechtel 2008, p. 153). This basically means that the causation occurring at different levels are mediated by the compositional/constitutive relation, and this accounts for the interaction between levels. In biological mechanisms such as the human brain, for example, there are thus, according to MTHC, many levels of constitutional/compositional organization, such as the level of connections between systems of neural networks, particular inter-cellular processes, intra-cellular processes, and molecular processes. In each of these constitutional/compositional levels, there are different causal processes always occurring at the same level, but these causal processes are mediated by the constitutional/compositional relations, affecting many different levels of the mechanism (cf. Craver and Bechtel 2007). There is here, therefore, causal pluralism and not reduction. Since (1) "whole systems exhibit behaviors that go beyond the behaviors of their parts" (Bechtel 2008, p. 129; cf. Craver 2007, p. 216), and since (2) the whole system is organized in a specific manner, and there are more causal processes occurring at this higher-level than the causal processes at a lower-level related to individual or sets of parts, then there is some independent causal higher-level—consequently causal plurality. In this sense, causation, according to this framework, is not an inter-level relation, but rather intra-level, i.e. it just occurs at the same level. Therefore, there is no upward (bottom-up) or downward (top-down) causation.

As a result, the mechanistic theory embraces a notion of causation which is not "fundamentalist" (cf. Craver 2007, p. 93). This means that there is no primary, lowest, or fundamental level (let us say the level of quarks or strings in physics) where the fundamental causal interactions that are the basis for the causal interactions in higher-levels take place. In Craver's view there is no "privileged level at which all causes act or at which all relevant causes are located" (2007, p. 104). And, according to the neo-mechanists, this causal pluralism is merely epistemological (explanatory) and does not lead to an ontological pluralism given the fact that new and mysterious causal interactions are not created completely from nothing, but rather are transmitted through the compositional mediated effects from the component parts to the whole higher-level mechanism.

Thus, what matters is that the interesting variables exhibit certain "patterns of manipulability" (Craver 2007, p. 104), i.e. that on each level it is possible to intervene on one variable to change other variables, to predict these changes and to control them. On this view, explanations "describe relationships between variables that can be exploited to produce, prevent, or alter" the phenomenon being explained (Craver 2007, pp. 100–101). That is, "X is causally relevant to Y if one can manipulate Y (or, more generally, the probability distribution over values of Y) by intervening ideally on X" and "X is explanatorily relevant to Y if it is causally relevant" (Craver 2007, p. 105). As long as there is the possibility of intervention and manipulation of mechanistic variables at a certain level, there are also for this reason genuine independent causal interactions at this level.

This also explains why no level has any kind of primacy or is more fundamental for the explanation and for the causal relations. In this sense, as Bechtel (2008, p. 148) points out "if we adopt the mechanistic account, [...] then we are not confronted with the prospect of a comprehensive lower level that is causally complete and closed".[11] Therefore, each level of investigation has its independence. The investigations at each level aim to understand "the causal processes" occurring at each level, which "are of different types" and "properly described in different vocabularies" (Bechtel 2008, p. 155). Therefore, what we have here is epistemological (explanatory) neuro-cognitive causal pluralism.

3.1.7 Neuro-Cognitive Pluralist Scientific Integration

The final major theoretical element in MTHC to be discussed concerns the idea that all these explanatory levels and causal processes, in spite of being autonomous in some sense, can be integrated in a pluralist (not reductionist) mechanistic scientific explanation (cf. Bechtel 2008; Boone and Piccinini 2015; Craver 2007; Kaplan 2017; Thagard 2019). The integration here has two dimensions: (1) an ontological, which refers to the phenomenon; and (2) an epistemological, which refers to the related scientific theories that explain this phenomenon. More particularly, the integration of the phenomenon refers to the idea that neuro-cognitive functions or capacities belong to natural biological multilevel neuro-cognitive mechanisms, and this is what must be completely understood; the integration of the theories refers to the idea that the theories concerned with the explanations of how these neuro-cognitive mechanisms work focus on these different levels of its organization and, thus, must be related in order to provide a complete explanation of the mechanism under consideration. I call this the mechanistic commitment to a pluralist scientific integration. In the case of mechanisms related to human cognition and brain, this concerns particularly the scientific fields and subfields interested in investigating the neuro-cognitive connection, e.g. experimental psychology, cognitive psychology, cognitive science, cognitive neuroscience and portions of neuroscience.

The basic idea is that even if there is some independence and autonomy between particular levels that compose a given mechanism, they need to be integrated in a comprehensive scientific explanation of the target mechanism as a whole. MTHC,

[11] For a discussion about this point cf. Kim (1998, 2005, 2009) and Papineau (2009).

therefore, aims at a double integration: (1) an integration between neural and cognitive mechanisms; and, at the same time, (2) an integration of the multiple scientific fields of inquire responsible for explaining these mechanisms and their functions (cf. Craver 2007, pp. 1, 2, 9, 16). This integration is implicit in the theory's appeal to an interrelated multilevel way of understanding mechanisms in neuroscience and cognitive science: the various levels are at the same time independent from each other and interrelated. As Bechtel synthetically points out, mechanistic explanation "requires integrating multiple levels of organization" (2008, p. 130).

Explanations of these mechanisms are provided by different fields and all of them contribute equally with constrains for the construction of the final explanation. All fields have their autonomy and are equally important. In his discussion of the mechanisms of spatial memory, Craver states the idea quite clearly: "Molecular biologists have the tools to design the knockouts", i.e. deleting a gene responsible for the production of some relevant protein, e.g. NMDA receptor; biochemists and histologists "have the tools to confirm the deletion"; the job of electrophysiologists is to "determine whether or not the knockout synapses can induce LTP"; systems-level electrophysiologists in turn "can monitor hundreds of cells at once to evaluate spatial map formation"; finally, "experimental psychologists are uniquely skilled at evaluating the spatial memory performance of the knockout mice", through systematic manipulation and observation of the behavior of the mice in a maze (Craver 2007, p. 265). Consequently, scientists working with these independent but connected levels should not work in isolation. They should rather communicate with each other. Working in collaboration they are more likely to find out all the aspects of the target mechanism, thereby being able to provide a more complete mechanistic account.

In this sense, the theory is compatible with scientific fields that also aim at such physicalist integration and provide also integrative scientific explanations. One of these scientific areas and indeed the one closer to this integrative ideal, in the view of neo-mechanists, is cognitive neuroscience, which emerged in the late 1980s aiming to present a scientific explanation for human cognition by investigating neural processes at different levels in the human brain (cf. Ochsner and Kosslyn 2014). The area is characterized as "the combined study of mind and brain" (Baars and Gage 2013, p. 3) and as a "bridging discipline between cognitive science and cognitive psychology, on the one hand, and biology and neuroscience, on the other" (Ward 2015, p. 2). More precisely, it is said that "the ultimate aim of cognitive neuroscience is to provide a brain-based account of cognition", i.e. to explain "how mental processes such as thoughts, memories and perceptions are organized and implemented by the brain"; or to put it in another way: "how the brain can create our mental world"; "how is it that a physical substance can give rise to our sensations, thoughts and emotions" (Ward 2015, pp. xi, 1, 2).

According to Bechtel, cognitive neuroscience is indeed a scientific field that integrates the disciplines of neuroscience and cognitive science and thereby integrates also "investigations of mind and brain" (2008, p. ix; cf. Boone and Piccinini 2015).[12]

[12]Boone and Piccinini (2015, Sect. 1) also argue more radically that traditional cognitive science "is being replaced by cognitive neuroscience". In their view, cognitive neuroscience constitutes a

As he points out, the mechanistic theory describes quite closely the actual scientific practice in cognitive neuroscience which is based on the identification and characterization of biological, neural and cognitive mechanisms (Bechtel 2009d, p. 20). The area assimilates in its explanatory strategy cognitive and behavioral tasks designed especially by experimental cognitive psychologists as well as neuroscientific methods for intervening on the brain and measuring brain activity. In Craver's view: "Cognitive neuroscience is, by its very nature, a field that encompasses psychological, physiological, cellular, and molecular items within its domain" (2007, p. 176). Thus, mechanistic multilevel explanations are the basis of the explanations developed by cognitive neuroscience. As Bechtel also points out, while one group of neuroscientists focuses on molecular and cellular activity, another group focuses on "systems, behavioral, and cognitive neuroscience" (2009d, p. 13). These investigators, he continues, "have appealed to systems-level understanding of the brain as providing the appropriate point of connection to the information processing accounts advanced in psychology" (2009d, p. 13).

The mechanistic approach "emphasizes the need to identify all (or at least the major) operating parts of the mechanism responsible for the phenomenon of interest and to understand the way they are organized and how their operations are orchestrated to realize the phenomenon" (Bechtel 2009d, p. 14). This demands a multilevel and integrative approach. In addition, the mechanistic theory is also not just concerned with a description of how neuroscientific and psychological research is done, but rather the framework also aims to give an account of how this kind of mechanistic explanations should be made. In this sense it gives also foundations and guidelines for the explanatory purpose of the disciplines interested in cognition and the brain. As Boone and Piccinini state:

> The scientific practices based on the old two-level view (functional/cognitive/computational vs. neural/mechanistic/implementation) are being replaced by scientific practices based on the view that there are *many* levels of mechanistic organization. No one level has a monopoly on cognition proper. Instead, different levels are more or less cognitive depending on their specific properties. The different levels and the disciplines that study them are not autonomous from one another. Instead, the different disciplines contribute to the common enterprise of constructing multilevel mechanistic explanations of cognitive phenomena. In other words, there is no longer any meaningful distinction between cognitive psychology and the relevant portions of neuroscience – they are merging to form cognitive neuroscience. (Boone and Piccinini 2015, Sect. 1)

These are the reasons why mechanistic explanations are not only compatible with those offered by cognitive neuroscience but they are also the most adequate for the explanatory and integrative needs and goals of this line of study, i.e. they can offer both foundations at a fundamental theoretical level and normative guidelines for the work done in this field.

However, it is also important to emphasize that the efforts of integration have already made a great influence in cognitive science and cognitive psychology. In present days, there is a great coverage of neuroscientific material in standard and

---

revolutionary break from tradition and "has emerged as the new mainstream approach to studying cognition".

popular textbooks of these areas (cf. this chapter, Sect. 3.1.1) and the mechanistic framework has become also influential inside them (cf. Anderson 2015; Bermúdez 2014; Eysenck and Keane 2015). As a result, MTHC can be as dominant in cognitive science and cognitive psychology as it is in cognitive neuroscience in case this integration continues to happen intensively and increasingly in the future. Indeed, MTHC tries to offer a major integrative theory to account for human cognition and for the human neuro-cognitive relation, both in the field of cognitive science and in the field of cognitive neuroscience.

3.2 MTHC Applied to Specific Cognitive Capacities

3.2.1 Perception

To illustrate more precisely and concretely how MTHC works in practice, it is useful to consider some examples. The first one is related to human perception, more particularly to the human visual system and the phenomenon of visual perception, which is basically understood as the capacity or ability of an organism (in this case humans) to acquire and process visual information from objects and events in the environment in order to be able to respond and act in a proper way (cf. Eysenck and Keane 2015, p. 35).

As a first step, the parts of the human brain related to the phenomenon of visual perception (or the capacity of visual perception) should be localized at least approximately as a way of starting the investigation, i.e. it is necessary to discover what parts of the human brain are more likely the components of the entire mechanism of visual perception. Once the multiple major brain areas related to visual perception are localized, the procedure of understanding their operations and organization can also take place. These brain areas arguably constitute the entire mechanism of visual perception in humans, i.e. these areas are where the parts of the mechanism are located. Since long time, it is common to recognize the occipital lobe as the center for vision. Many studies that have been conducted show deficits in visual processing due to damage in the occipital lobe—this is true for animals from a variety of species, including humans. Such results are also supported by neuroanatomical studies which show a projection of the optic tract going from the eye, passing by the *lateral geniculate nucleus* (LGN), which is an area of the thalamus, and achieving the occipital lobe (Bechtel 2008, p. 92). The model of the mechanism under investigation here would thus include the eyes and optic nerves and the brain areas responsible for visual perception, namely the occipital lobe and areas in temporal and parietal lobes (more precisely, the posterior parietal cortex and the inferotemporal cortex), as current research indicates (Bechtel 2008, chap. 3; cf. Eysenck and Keane 2015, p. 37).

The next step is to discover what operations each part of the mechanism performs. The occipital lobe, for instance, can be itself decomposed in areas responsible for visual perception, such as the *striate cortex*, also known as *Brodmann area 17*, or

V1 (primary visual cortex, or visual area 1). Each of these components need to be understood. For example, what are the operations localized in *V1*? The same procedure needs to be done for all the other areas in the brain which are also part of the mechanism responsible for visual perception; for instance, *V2*, *V3*, *V4*, *V5/MT*. The goal is to discover which cells (including visual receptor cells in the retina of the eye, such as cones and rods), small networks of cells, or larger neural systems in these areas are responsible for processing information about light and dark spots, bars of light (edges), size, shape, color, depth, location and motion of objects in the visual field. Moreover, it is also important to discover the pathways and channels through which the information is transmitted.

Currently, however, the operations of many areas related to visual processing remain unknown, since the whole mechanism is enormously complex and, therefore, the evidence is insufficient to determine all of them (Bechtel 2008, p. 127). More research is necessary in order to obtain a more complete picture of the overall mechanism. At any rate, even if far from complete, according to Bechtel (2008, p. 127), "the account of the visual processing mechanism is one of the best worked out accounts of a mental mechanism and a worthy exemplar".

3.2.2 Memory

The second example concerning the application of MTHC to a concrete human cognitive capacity is related to memory. Functional analyses of the memory capacity reveal the existence of many memory sub-capacities; i.e. the general memory function can be subdivided in many particular sub-functions. Current research indicates, for example, the existence of short-term memory, working memory, long-term memory, visual memory, auditory memory, semantic memory, episodic memory, etc., as well as different processes related to them, such as encoding, storage, retention, and retrieval (cf. Bechtel 2008, chap. 2; Eysenck and Keane 2015, chap. 6). In a mechanistic explanation, this is the procedure called functional decomposition (cf. Chap. 2, Sect. 2.1.5). This procedure shows that the memory system is composed by many subsystems (or sub-functions). Due to this major complexity within the memory system, it is more useful to seek explanations for each sub-function first, and then try to understand how they are related considering the whole mechanism of memory.

One of the best understood phenomena in the memory system so far is memory consolidation. Roughly put, the process of memory consolidation stabilizes a memory after its initial acquisition, i.e. it is the phenomenon of transforming short-term memories into long-term memories. The central idea behind this process of memory consolidation, thus, is that, for a period of time, new memories are labile and easy to disrupt, but when consolidation takes place, they become robust and enduring (Bechtel 2009d,

p. 15; 2008, chap. 2).[13] This in turn permits the organism to remember events for a longer period of time and modify its behavior accordingly when necessary. For example, if a mouse is able to remember some negative consequence of being in a certain place, it is more likely that it will avoid going to that place again in the future. If the mouse can remember a positive consequence of being in a certain place, it will likely seek to go to that same place more frequently (cf. e.g. the water maze experiments with mice).

Three processes related to memory consolidation are often distinguished in the literature: (1) synaptic consolidation, that occurs within the first few hours after learning; (2) systems consolidation, in which memories become independent of the hippocampus (being stored in different areas of the neocortex), and occurs over a period of weeks to years after the memory is acquired; and (3) reconsolidation. For our purposes here, it is enough to consider just the process related to (1), synaptic consolidation.

To explain this phenomenon, all the relevant regions in the brain responsible for the functions that compose this neuro-cognitive mechanism of memory consolidation, including all relevant activities performed by these component parts must be determined. In addition, the causal processes and causal interactions within the mechanism functions need also to be understood, i.e. the general organization of the mechanism. Accordingly, the whole mechanism must be decomposed in its working parts and organization must be understood. Besides, all the activities and causal interactions need to be related to real neural activities though the process of localization. In order to achieve this goal, it is important to relate all the different mechanistic levels relevant for the explanation of the phenomenon.

Firstly, there is the neural systems level of "the hippocampus" and its "neuro-architecture" (Bechtel 2009d, pp. 16, 22). The hippocampus is arguably formed by the neural structures CA_1, CA_2, CA_3, CA_4 and the *Dentate Gyrus*, and it is localized in the medial-temporal lobe in human brains. Surrounding it there are the important regions of the *pirhinal cortex*, *parahippocampal cortex* and *entorhinal cortex*. One of the reasons for the belief that the hippocampus plays a very important role in memory consolidation is the fact that the removal or damage in this neural structure leads to impairment in particular memory capacities, in humans and other nonhuman animals. The most compelling evidence so far regarding this is the famous case of H.M. (Craver 2005, p. 378). Having his life threatened by episodes of epileptic seizures, H.M. decided to undergo medical surgery which removed the hippocampus from both sides of his brain. After the surgery he could write and read, learn new skills, remember much of his childhood; so, his procedural memory (for learning new skills) was not impaired. However, he could not remember particular events (episodic memory) that happened during a given period of his life for 11 years before the removal of his hippocampus (retrograde amnesia), nor recent events after the removal (anterograde amnesia). Therefore, it is also believed that the hippocampus

[13]Craver (2007, chap. 5; cf. 2002) offers the example of spatial memory to illustrate mechanistic multilevel explanations. However, his considerations are very similar to these of memory consolidation.

is especially related to episodic memory consolidation and not all kinds of memory consolidation.

The large neural network present in the hippocampus can be thus considered the whole mechanism of episodic memory consolidation, since one of the functions (among others) that this whole mechanism performs is the production of the phenomenon of episodic memory consolidation, which is the *explanandum*. It is this capacity that makes it possible for the organisms, including humans, to behave in particular way; for instance, to remember a given fact and act accordingly. A good explanation needs to include a great amount of understanding at this highest level of the whole mechanism. At this level, it is necessary to correctly identify all the large neural structure that is responsible for the phenomenon and to understand if this large neural system is indeed all that is relevant for the explanation of the phenomenon. For this, techniques such as fMRI can be useful, since they can point out what regions in the brain are activated during the performance of a specific psychological task. The mechanistic explanation at this level also needs to clarify how the neural network encode and decode new memory episodes through information processing and computational operations and how these processes produce and affect, for instance, the different degrees of consolidation that characterize the memories under investigation, thereby influencing the organism's behavior. This large neural system needs to be decomposed into particular sub-neural systems for the understanding of the whole network. The goal of the explanation at this level is to understand the information processing and computational operations of the neural networks and how they contribute to the performance of the whole mechanism composed by such neural-nets. At this particular level, the components of a particular neural network need to be correctly understood. Component parts are a small number of neurons and their operations of, for example, depolarizing and firing, in the process of propagation of action potentials. Moreover, they are responsible for synaptic processes, neurotransmitters being released, and so on. Here it is possible to measure spiking rates of neurons, or spiking duration.

The explanation can be extended to another level of decomposition: the intercellular, intracellular and molecular level. At this level, the description is in terms of the activity of relevant particular neurons, synapses, neurotransmitters, receptors, proteins, molecules and ions. For instance, it has long been suggested that consolidation of memories is a process that is strongly related to the enhancement in the activity of synapses of neurons in the brain. This process has been called long-term potentiation (LTP), which many authors think is a crucial process for memory consolidation. According to the mechanists, LTP is a component part in the hippocampus mechanism for episodic memory consolidation, together with various other synaptic components (cf. Craver 2002, p. 89). Although many aspects related to this process remain unclear, there is already evidence to clarify in general lines how it might work according to some proposed models. The process is considered a form of persistent synaptic plasticity (since it lasts from several minutes to many months), which is the capacity of synapses to change their strength. It is understood as a persisting strengthening (long-last increasing in signal transmission) of synapses in the response to recent patterns of activity between two neurons (the presynaptic and the postsynaptic when simultaneously active) (cf. Bechtel 2009d, p. 16; Craver 2002, p. 85ff). Many authors consider this process to be central for the consolidation of memories at the purely physiological level. They argue that memories can be encoded in the brain through the modification of synaptic strength. Research on LTP is typically made in the hippocampus, where it occurs in particular cells localized in particular regions, which given their specific neat organization, are easier to be investigated. However, it has also been observed in some other neural structures (e.g. amygdala and cerebral cortex). At any rate, hippocampal-LTP can be considered a component in the hippocampus for episodic memory consolidation (cf. Craver 2005, p. 389).

There are two phases of the process of LTP: early LTP (E-LTP), which does not require gene transcription and protein synthesis; and late LTP (L-LTP) which does require gene transcription and protein synthesis. There are also two phases of L-LTP: phase 1, which requires protein synthesis, and phase 2, which requires protein synthesis and also gene transcription. These phases are often named in the literature as LTP_1 (for E-LTP), LTP_2 (for L-LTP phase 1), and LTP_3 (for L-LTP phase 2). The components of their processes are characterized roughly as follows.

In E-LTP, calcium-calmodulin kinase II (CaMKII) and protein kinase C (PKC) phosphorylate existing α-amino-3-hydroxy-5-methyl-4-isoxazole proprionic acid (AMPA) receptors to increase their activity and mediate or modulate the insertion of additional AMPA receptors into the postsynaptic membrane, without any protein synthesis. By increasing the efficiency and number of AMPA receptors at the synapse, future excitatory stimuli generate larger post-synaptic responses.

In L-LTP, when the pre-synaptic neuron is activated it releases the neurotransmitter glutamate that binds to N-methyl-D-aspartate (NMDA). NMDA, in turn, changes its shape and exposes channels (pores) in the membrane of the post-synaptic neuron. In case the post-synaptic neuron is in its polarized resting state, the channels remain blocked by magnesium ions (Mg^{2+}), but in case the neuron is depolarized, the Mg^{2+} ions leave the channel and allow calcium ions (Ca^{2+}) enter the cell and initiate chemical reactions that alters the properties of AMPA receptors. These receptors connect with glutamate as well, but they regulate in addition the flow of sodium ions (Na^+) and potassium ions (K^+) in the post-synaptic cell. All these chemical events regulate what is required for L-LTP. In this way, the NMDA receptors acting as gates ensure that LTP takes place only when the pre- and post-synaptic cells are simultaneously active. Moreover, there is also a process protein synthesis through the activation of the DNA in the post-synaptic neuron nucleus. Cyclic adenosine monophosphate (cAMP) and other molecules activate protein kinase A (PKA), which communicates with the nucleus. In the nucleus, PKA phosphorylate cAMP response element binding protein (CREB), which then binds with DNA and activates the transcription of genes that initiate synthesis of two proteins. The first suppresses the activity of PKA and the other facilitates an increase in the number of active receptors in the post-synaptic cell (cf. Bechtel 2009d, p. 18).

These proteins and molecules are considered to be important for the processes of LTP, i.e. components in the mechanism of LTP, which is a sub-sub-subsystem in the whole mechanism of memory consolidation and play a role for the phenomenon to take place, together with other operations. These other operations, however, are not clarified by the neo-mechanists.

Arguably, this kind of explanation exhibits different levels for the characterization of memory consolidation (Bechtel 2009d, p. 18; cf. Bechtel 2008, chap. 3; Craver 2007, chap. 5). All levels are equally important to achieve the complete multilevel mechanistic explanation of the particular phenomenon in the end, since in all of the higher-levels there are arguably causal processes that are not found in lower-levels.[14]

This is a paradigmatic example of MTHC applied to a concrete human cognitive phenomenon. However, as we can see, there are still many details that cannot yet be fulfilled, due to lack of information and clear evidence. There is indeed already many information about the molecular level, in which many structures are believed to play a relevant role for memory consolidation in certain cognitive biological systems. Nevertheless, when the higher neural mechanistic levels are considered, it is not well understood how exactly they perform the functions that some authors believe they perform; for example, how exactly the hippocampus 'compute' information. And what about the content of the representation that is being memorized? It appears that while we understand what are the main neural mechanisms involved in episodic long-term memory, the account tells us almost nothing about the content being memorized and how it is memorized.

At any rate, it is possible to observe that in these examples of mechanistic psychological phenomena two generally different strategies for building mechanistic explanations of cognitive phenomena were implemented. In the case of episodic memory consolidation, initial attention was given to functional decomposition and a better understanding of the particular phenomenon being explained. In the case of visual perception, the initial strategy was to discover the brain regions that are related to the phenomenon and just afterwards pay more attention to the operations they perform. This difference is due to the possibility of a more direct access of the responsible regions in the brain, since it is easier to present stimuli to the sensory organs like the eye and track its effects on brain structures and pathways moving through different areas according to the stimulation effects.

These two descriptions of how the mechanistic theory of explanation works in actual, concrete cognitive science (applied to episodic memory consolidation and visual perception) show that it is a particular form of constructing scientific explanations. Particularly, these examples show that mechanistic explanations are fundamentally different from explanations formulated according to classic DNTSE. MTHC, grounded in MTCBS, contrarily to DNTSE based on physics, does not focus on logical deductions and general laws, which are typically, in the view of the neo-mechanists, not of interest for the concrete work of neuroscientists and cognitive scientists. No step of the explanations provided above contains general laws of nature and particular initial conditions. No deductive argument was provided, and no attempt to provide one was made. Nor there was any inductive argument where its consequence follows with high probability.

---

[14]I will discuss this point deeper in the next chapter.

This is because the mechanistic framework focuses on the internal works of natural mechanisms, their components, operations and organization. A mechanistic explanation must show how the natural mechanism under consideration produces a given phenomenon (performs a given function). This explanation is provided by understanding the internal works of this mechanism, i.e. its component parts and organization, as well as by understanding how external causal factors influence the activities of the particular mechanism. Thus, scientific explanation in cognitive science is not a matter of constructing general laws of nature and constructing deductive or inductive arguments, as defended by the classical account. The point is rather to provide an accurate description of how the causal processes and causal interactions taking place inside a cognitive biological mechanism explains a particular function it performs. Mechanistic explanations in cognitive science are causal explanations, but generalizations need to be made with caution, since generalizations in the biological sciences are fragile, i.e. they can be limited, for example, to a given population of organisms, or to a particular point in time, and to other particular conditions. Different mechanistic structures and their functions need to be compared in detail so that generalizations can be correctly established.

3.3 Final Remarks

I presented, in this chapter, one of the most influential scientific and philosophical fundamental theories of human cognition in the twenty-first century: MTHC. This theory is explicitly or implicitly embraced by many authors in cognitive neuroscience, cognitive science, and particular fields of philosophy. The theory's central theoretical elements were systematically presented and discussed together with its most fundamental goals.

It is not a simple task, however, to understand and present the most central theoretical elements of MTHC. Many of these elements are still highly debated and need further detailed elaboration. Furthermore, the theory is highly broad and ambitious: it takes notions, concepts and ideas from a variety of fields, such as analytical philosophy of mind, analytical philosophy of science, analytical philosophy of cognitive science, cognitive science, cognitive neuroscience, neuroscience and fields of biology. Here I tried to offer an accurate and plausible formulation of this theory, combining the work of some of its most prominent advocates. Primarily, I based this construction of the mechanistic theory of human cognition in the work of William Bechtel, who presents the most detailed discussion of the theory so far. However, I considered the work of Carl Craver to a great extent as well, since he also provides detailed discussion on these matters. Moreover, I considered the works of Gualtiero Piccinini, Paul Thagard, and some other influential neo-mechanists working with human cognition and cognitive science. Thus, the version of this theory that was presented here is not the one of a specific author, but it is rather an integrated perspective on the MTHC, which is particularly suitable to be reasonably and systematically evaluated without distortions or omissions.

The theory, based on a physicalist background, has been establishing foundations and goals for a large part of contemporary scientific and philosophical work on human cognition, especially since the final part of the twentieth century. MTHC views human cognitive processes as neural computations which represent and process information coming from external stimuli. A mechanistic explanation clarifies in what exactly such computations consist and where exactly they take place in the brain. One of the most important aspects of the theory thus is that it attempts to integrate explanations of cognitive and neural phenomena in a coherent framework. Accordingly, neo-mechanists argue for the strategy of identifying types of cognitive functions with types of concrete physio-chemical neural processes taking place in brains. Such neural processes are performed by neural mechanisms that can be fully decomposed in many levels of organization. Moreover, since the theory pleas for integration of multilevel mechanisms and multiple fields of research, establishing brain research as one of the most important areas of investigation, it is especially compatible with the explanatory aims of cognitive neuroscience, but also with a research program in cognitive science in which neuroscience plays a central role.

The attempt to integrate the sciences interested in human cognition by offering a comprehensive theory of human cognition is important and perhaps even necessary. However, it is at the same time a very difficult task and an extremely ambitious project, given the extent of diversity and fragmentation in the relevant fields. With the analysis offered in this chapter the reader can note that, despite of the current popularity of the mechanistic point of view, MTHC faces a great amount of difficulties concerning its very formulation. There are still many terminological disputes related to many of its central concepts and it is not entirely clear to what extent these different terms carry or not different meanings.

There are as well internal disputes of major proponents concerning central theoretical aspects of MTHC, e.g. the role identity relations play in the theory, or how exactly to characterize cognitive/mental mechanisms. The fact that there are "salient differences between the various accounts of mechanism" has been recognized by its own advocates (cf. Bechtel and Abrahamsen 2005, p. 423). This makes it very hard to identify a common structure and formulation of the framework. Moreover, many aspects of the theory are not yet developed with major accuracy. For instance, what sort of computation biological cognitive mechanisms perform and how exactly such computations can be used to integrate processes occurring at a biological neural level and processes occurring at a higher cognitive level of human reasoning and consciousness? The theory does not have a clear answer for this question.

To sum up, a great amount of theoretical integrative work concerning human cognition and its explanation by the fields of cognitive science and cognitive neuroscience has been done, and this deserves praise. However, MTHC still needs further developments concerning central theoretical elements and particular theoretical elements, if it is to be proposed indeed as a comprehensive and integrative theory of human cognition. This point is accepted even by its own advocates: "The new mechanical philosophy [...] has expanded rapidly over the last two decades", however, "it is clear that many of the major topics are only beginning to develop, leaving a lot of work for scholars to elaborate the basic commitments of this framework" (Craver and Tabery 2015, Sect. 7).

References

Anderson, J.L.: How Can the Human Mind Occur in the Physical Universe?. Oxford University Press, Oxford (2007)

Anderson, J.R.: Cognitive Psychology and its Implications, 8th edn. Worth Publishers, New York (2015)

Baars, B.J., Gage, N.M.: Fundamentals of Cognitive Neuroscience. Elsevier, Oxford, UK (2013)

Bechtel, W., Abrahamsen, A.: Explanation: A mechanist alternative. Stud. Hist. Philos. Biol. Biomed. Sci. **36**, 421–441 (2005)

Bechtel, W., McCauley, R.N. Heuristic identity theory (or back to the future): The mind-body problem against the background of research strategies in cognitive neuroscience. In Hahn, M., Stoness, S.C. (eds.) Proceedings of the 21st Annual Meeting of the Cognitive Science Society, pp. 67–72. Lawrence Erlbaum Associates, Mahwah, NJ (1999)

Bechtel, W., Wright, C.D.: What is psychological explanation? In: Symons, John, Calvo, Paco (eds.) The Routledge Companion to Philosophy of Psychology, pp. 113–130. Routledge/Taylor & Francis Group, New York (2009)

Bechtel, W.: Decomposing the mind-brain: A long-term pursuit. Brain and Mind **3**, 229–242 (2002)

Bechtel, W.: Mental Mechanisms: Philosophical Perspectives on Cognitive Neurosciences. Routledge, New York (2008)

Bechtel, W.: Constructing a philosophy of science of cognitive science. Top. Cogn. Sci. **1**(3), 548–569 (2009a)

Bechtel, W.: Looking down, around and up: Mechanistic explanations in psychology. Philos. Psychol. **22**(5), 543–564 (2009b)

Bechtel, W.: Mechanism, modularity, and situated cognition. In: Robbins, Philip, Aydede, Murat (eds.) The Cambridge Handbook of Situated Cognition, pp. 155–170. Cambridge University Press, Cambridge (2009c)

Bechtel, W.: Molecules, systems, and behavior: Another view of memory consolidation. In: Bickle, J. (ed.) The Oxford Handbook of Philosophy and Neuroscience, pp. 13–40. Oxford University Press, New York (2009d)

Bechtel, W.: How can philosophy be a true cognitive science discipline? Top. Cogn. Sci. **2**, 357–366 (2010)

Bechtel, W.: Investigating neural representations: The tale of place cells. Synthese **193**, 1287–1321 (2016)

Bermúdez, J.L.: Cognitive Science: An Introduction to the Science of the Mind, 2nd edn. Cambridge University Press, Cambridge (2014)

Boone, W., Piccinini, G.: Cognitive neuroscience revolution. Synthese (2015). https://doi.org/10.1007/s11229-015-0783-4

Churchland, P.S.: Neurophilosophy. MIT Press, Cambridge, MA (1986)

Craver, C.F., Bechtel, W.: Top-down causation without top-down causes. Biol. Philos. **22**, 547–563 (2007)

Craver, C.F., Tabery, J. Mechanisms in science. In: Zalta, E.N. (ed.) The Stanford Encyclopedia of Philosophy, Spring 2017 edn. https://plato.stanford.edu/archives/spr2017/entries/science-mechanisms/ (2015)

Craver, C.F.: Interlevel experiments and multilevel mechanisms in the neuroscience of memory. Philos. Sci. **69**(S3), S93–S97 (2002)

Craver, C.F.: Beyond reduction: Mechanisms, multifield integration and the unity of neuroscience. Stud. Hist. Philos. Biol. Biomed. Sci. **36**, 373–395 (2005)

Craver, C.F.: Explaining the Brain: Mechanisms and the Mosaic Unity of Neuroscience. Oxford University Press, Oxford (2007)

Cummins, R.: "How does it work?" versus "what are the laws?": Two conceptions of psychological explanation. In: Keil, F., Wilson, R.A. (eds.) Explanation and Cognition, pp. 117–145. MIT Press, Cambridge, MA (2000)

Eliasmith, C.: How to Build a Brain: A Neural Architecture for Biological Cognition. Oxford University Press, Oxford (2013)

Eysenck, M.W., Keane, M.T.: Cognitive Psychology: A Student's Handbook, 7th edn. Psychology Press, New York (2015)

Feigl, H. The "Mental" and the "Physical". In: Feigl, H., Scriven, M., Maxwell, G. (eds.) *Concepts, Theories and the Mind-Body Problem* (Minnesota Studies in the Philosophy of Science, vol. 2). University of Minnesota Press, Minneapolis (1958)

Frankish, K., Ramsey, W.: The Cambridge Handbook of Cognitive Science. Cambridge University Press, Cambridge (2012)

Gardner, H.: The Mind's New Science: A History of the Cognitive Revolution. Basic Books, New York (1985)

Gazzaniga, M.S., Ivry, R.B., Mangun, G.R.: Cognitive Neuroscience: The Biology of the Mind, 4th edn. W. W. Norton, New York (2014)

Goldstein, B.: Cognitive Psychology: Connecting Mind, Research and Everyday Experience, 4th edn. Cengage Learning, Stamford, CT (2015)

Gozzano, S., Hill, C.S. (eds.): New Perspectives on Type Identity: The Mental and the Physical. Cambridge University Press, Cambridge (2012)

Harnish, R.M.: Minds, Brains, Computers: An Historical Introduction to the Foundations of Cognitive Science. Blackwell Publishers, Malden (2002)

Kaplan, D.M. (ed.): Explanation and Integration in Mind and Brain Science. Oxford University Press, Oxford (2017)

Kim, J.: Mind in a Physical World: An Essay on the Mind-Body Problem and Mental Causation. MIT Press, Cambridge, MA (1998)

Kim, J.: Physicalism, or Something Near Enough. Princeton University Press, Princeton (2005)

Kim, J.: Mental causation. In: McLaughlin, B.P., Beckermann, A., Walter, S. (eds.) The Oxford Handbook of Philosophy of Mind, pp. 29–52. Oxford University Press, Oxford (2009)

Laird, J.E.: The Soar Cognitive Architecture. MIT Press, Cambridge, MA (2012)

Machamer, P., Darden, L., Craver, C.: Thinking about mechanisms. Philos. Sci. **67**, 1–25 (2000)

McCauley, R.N., Bechtel, W.: Explanatory pluralism and heuristic identity theory. Theor. Psychol. **11**(6), 736–760 (2001)

McCauley, R.N.: About face: Philosophical naturalism, the heuristic identity theory, and recent findings about prosopagnosia. In: Gozzano, S., Hill, C.S. (eds.) New Perspectives on Type Identity: The Mental and the Physical, pp. 186–206. Cambridge University Press, Cambridge (2012)

Miller, G.A.: The cognitive revolution: A historical perspective. Trends Cogn. Sci. **7**(3), 141–144 (2003)

Mnih, V., Kavukcuoglu, K., Silver, D., Rusu, A.A., Veness, J., Bellemare, M.C., Hassabis, D.: Human-level control through deep reinforcement learning. Nature **518**, 529–533 (2015)

Neisser, U. Cognitive Psychology. Psychology Press, New York. (2014, Originally published in 1967)

Ochsner, K.N., Kosslyn, S.M. (eds.): The Oxford Handbook of Cognitive Neuroscience, vol. 1. Oxford University Press, Oxford (2014)

Oppenheim, P., Putnam, H.: The unity of science as a working hypothesis. In: Maxwell, G., Feigl, H., Scriven, M. (eds.) Concepts, Theories, and the Mind-Body Problem, pp. 3–36. Minnesota University Press, Minneapolis (1958)

Papineau, D.: The rise of physicalism. In: Gillett, C., Loewer, B. (eds.) Physicalism and Its Discontents, pp. 3–36. Cambridge University Press, Cambridge (2001)

Papineau, D.: The causal closure of the physical and naturalism. In: McLaughlin, B.P., Beckermann, A., Walter, S. (eds.) The Oxford Handbook of Philosophy of Mind, pp. 53–65. Oxford University Press, Oxford (2009)

Papineau, D. Naturalism. In: Zalta, E.N. (ed.) The Stanford Encyclopedia of Philosophy, Winter 2016 edn. https://plato.stanford.edu/archives/win2016/entries/naturalism/ (2015)

Piccinini, G., Craver, C.: Integrating psychology and neuroscience: Functional analyses as mechanism sketches. Synthese **183**(3), 283–311 (2011)

Piccinini, G.: Computationalism. In: Margolis, E., Samuels, R., Stich, S. (eds.) The Oxford Handbook of Philosophy of Cognitive Science, pp. 222–249. Oxford University Press, New York (2012)

Place, U.T.: Is consciousness a brain process? Br. J. Psychol. **47**, 44–50 (1956)

Rosenbloom, P.S., Demski, A., Ustun, V.: The sigma cognitive architecture and system: Towards functionally elegant grand unification. J. Artif. Gen. Intell. **7**, 1–103 (2016)

Smart, J.: Sensations and brain processes. Philos. Rev. **68**(2), 141–156 (1959)

Smith, E.E., Kosslyn, S.M.: Cognitive Psychology: Mind and Brain, International edn. Pearson, Edinburgh (2014)

Sternberg, R.J., Sternberg, K.: Cognitive Psychology, 6th edn. Wadsworth, Cengage Learning, Belmont, CA (2012)

Stoljar, D.: Physicalism. Routledge, New York (2010)

Stoljar, D. Physicalism. In: Zalta, E.N. (ed.) The Stanford Encyclopedia of Philosophy, Spring 2015 edn. http://plato.stanford.edu/archives/spr2015/entries/physicalism/ (2015)

Sturm, T., Gundlach, H. Zur Geschichte und Geschichtsschreibung der 'kognitiven Revolution' – eine Reflexion. In: Stephan, A., Walter, S. (Hrg.) Handbuch Kognitionswissenschaft, pp. 7–22. Metzler, Stuttgart (2013)

Thagard, P.: Mind: Introduction to Cognitive Science, 2nd edn. MIT Press, Cambridge, MA (2005)

Thagard, P.: Hot Thought: Mechanisms and Applications of Emotional Cognition. MIT Press, Cambridge, MA (2006)

Thagard, P.: Why cognitive science needs philosophy and vice versa. Top. Cogn. Sci. **1**, 237–254 (2009)

Thagard, P. Cognitive science. In: Zalta, E.N. (ed.) The Stanford Encyclopedia of Philosophy, Winter 2018 edn. https://plato.stanford.edu/archives/win2018/entries/cognitive-science/ (2018)

Thagard, P. Brain-Mind: From Neurons to Consciousness and Creativity (Treatise on Mind and Society). Oxford University Press, Oxford (2019)

Von Eckard, B.: What Is Cognitive Science?. MIT Press, Cambridge, MA (1993)

Ward, J.: The Student's Guide to Cognitive Neuroscience, 3rd edn. Psychology Press, New York (2015)

Wright, C., Bechtel, W.: Mechanisms and psychological explanation. In: Thagard, Paul (ed.) Philosophy of Psychology and Cognitive Science, pp. 31–79. North Holland Elsevier, Amsterdam (2007)

Zednik, C.: Mechanisms in cognitive science. In: Glennan, S., Illari, P. (eds.) The Routledge Handbook of Mechanisms and Mechanical Philosophy, pp. 389–400. Routledge, London (2018)

Part II
Major Challenges to the Mechanistic Theory of Human Cognition

Chapter 4
Molecular and Cellular Theory of Human Cognition

4.1 MCTHC Against MTHC

4.1.1 The Issue of Neuro-Cognitive Reduction

Roughly, a neuro-cognitive reductionist thesis states that all cognitive phenomena (including all the phenomena related to consciousness and self-consciousness) can be 'reduced' to neural phenomena. In this sense, human neural processes ultimately 'produce' human cognitive processes. At the same time, a complete explanation of human neural processes provides an explanation of human cognitive processes. The mechanistic theory wants to 'integrate' human cognitive phenomena and explanations with a neuroscientific explanatory framework in a 'pluralist' way (cf. Chap. 3, Sect. 3.1.7). As a result, the question of reduction immediately arises: after all, are mechanistic multilevel explanations also in some sense reductive explanations?

Reductive approaches in science and philosophy are attractive for at least two reasons. The first one is ontological, i.e. related with the very phenomena happening in nature. Reductions of phenomena give rise to a simpler and more integrated (unitary) conception of the phenomena in reality. For example, if all mental phenomena can be reduced to physical phenomena, they can be simply assimilated by the currently generally accepted (especially in the natural sciences and portions of analytic philosophy) ontology of the world, which is essentially physicalist (cf. Kim 1998, 2005). In this sense, one could say, for instance, that the cognitive phenomenon C_p is reduced to the neural phenomenon N_p (e.g. the phenomenon of human sensation of pain is reduced to some pattern of neural connection in the limbic system). Consequently, there would be just the physical realm and all phenomena would be explained in a simpler and more unified manner on the same general basis, i.e. physically.

Secondly, reductive approaches are attractive because they have also implications for science, especially for the relationship between scientific theories and for the relationship between scientific fields. It can be argued that reductions help in the advancement of the organization of human knowledge, integrating and unifying it.

D. A. Leite, *The Twenty-First Century Mechanistic Theory of Human Cognition*, Cognitive Systems Monographs 41, https://doi.org/10.1007/978-3-030-63680-7_4

Reductions were originally employed in order to help achieving "the ideal of a comprehensive theory which will integrate all domains of natural science in terms of a common set of principles and will serve as the foundation for all less inclusive theories" (Nagel 1961, p. 336). The phenomenon of reduction, in which "a relatively autonomous theory" becomes "absorbed by, or reduced to, some other more inclusive theory" was seen as "an undeniable and recurrent feature of the history of modern science", and one would have "every reason to suppose that such reduction will continue to take place in the future" (Nagel 1961, pp. 336–337).

The implications of reduction for science concern theories or domains which can be reduced to more fundamental theories or domains. For instance, one could claim that the cognitive theory C_t could be reduced to the neurophysiological theory N_t (e.g. some psychological theory about human visual perception could be completely reduced to a neurophysiological theory about patterns of neural connection in the occipital lobe; or that a theory about the nature of human pain sensation can be reduced to some neural theory concerning patterns of connection in the limbic system).[1] Intertheoretical scientific reductions arguably can help to avoid gaps between scientific terms and theories and also avoid their multiplication beyond necessity. Indeed, a natural consequence of scientific progress and specialization is a certain proliferation of terms and theories. However, in certain cases, different vocabularies or theories in different domains (e.g. chemistry and biology; psychology and neuroscience) can be used to explain the same phenomena. If the explanatory power of one of these theories is stronger, this theory can arguably reductively integrate the other into a unified theory.

Besides these more 'theoretical implications' that the issue of reduction brings to the sciences, there are other more 'pragmatic implications', as well. These latter issues concern, for example, "debates over what is the most promising direction systematic research should take at some given stage of a science" (Nagel 1961, p. 363). In the particular case of cognitive science and cognitive neuroscience, one can ask: what is the best research strategy to inquire about human cognition? Which of these disciplines should receive more financial investment? The administrative staff of a given university should open a course on cognitive science or cognitive neuroscience for the inquiry about human cognition? What about laboratories, research groups, equipment, etc.? All these questions and the respective debates over them are related to issues concerning the relationship between the sciences and the best research strategy to follow, i.e. debates concerning the integration and improvement of the sciences, also in a more concrete way. Therefore, philosophical and scientific accounts of reduction, as a possible rigorous form of improving human knowledge and scientific integration and development, can provide theoretical and pragmatical guidelines. Thus, it is highly important.

---

[1] There is indeed a strict relation between the ontological dimension of reduction and its intertheoretical dimension. The literature on this relationship though is deeply controversial. In the present chapter I will focus on issues more related to epistemology, but I will address issues of ontology in Chap. 6, Sect. 6.1.3.

In this context, the more particular hypothesis that human cognitive capacities are somehow 'dependent on' physio-chemical neural processes is our primary concern. The general idea is that a scientific objective and accurate explanation of the brain operations will show that neural processes somehow are the 'basis of' cognitive capacities. This 'reductive' idea was already grounding rigorous and methodic scientific works in the middle of the twentieth century, such as Donald Hebb's *Organization of Behavior* (1949), where he provides a theory concerning the neural basis of learning behavior; or the work of McCulloch and Walter Pitts on neural networks, *A Logical Calculus of Ideas Immanent in Nervous Activity* (1943), where the authors identify the cognitive processing of propositions with neural activity. Many scientific works, indeed, endorse some kind of reductionist view (e.g. Edelman 1989, 2004; Ramachandran 2011; Soon et al. 2008).

One of the best examples of this way of thinking is found in the work of Francis Crick, *The Astonishing Hypothesis* (1994). The main thesis of the author is that "'You', your joys and your sorrows, your memories and your ambitions, your sense of personal identity and free will, are in fact no more than the behavior of a vast assembly of nerve cells and their associated molecules" (Crick 1995, p. 3). In his view: "The scientific belief is that our minds – the behavior of our brains – can be explained by interactions of nerve cells (and other cells) and the molecules associated with them" (Crick 1995, p. 7). Moreover, he explicitly assumes a reductionist perspective on the issue: "[…] 'reductionism' is the main theoretical method that has driven the development of physics, chemistry, and molecular biology" (Crick 1995, p. 8).

Another clear example is found through the works of Patricia Churchland. The author clearly assumes that neural activity is responsible for creating mental life: "To understand how neurons give rise to a mental life, we must know what they do, both individually as single cells and collectively as coherent systems of cells" (Churchland and Sejnowski 1992, p. vi). She assumes that "neuroscience can reveal the physical mechanisms subserving psychological functions", since "it is indeed the brain that performs those functions" and "capacities of the humans mind are in fact capacities of the human brain"; and she is "convinced that the right strategy for understanding psychological capacities is essentially reductionist", which means "trying to explain the macro levels (psychological properties) in terms of micro levels (neural network properties)" (Churchland 1994, p. 23).

Since the notion of reduction has some important attractive aspects and has been playing a central role in theorizing about the relationship between human brain and cognition, it is certainly important to precisely investigate whether the thesis of neurocognitive reduction is plausible or not. In order to do this, though, it is necessary to understand correctly the underlying theory of reduction, i.e. what reduction really means. According to van Riel and Van Gulick (2014, intro.; cf. Van Gulick 2001), to claim that the mental reduces to the physical, that heat reduces to kinetic molecular energy, or that one theory reduces to another theory, is to imply metaphorically that in "some relevant sense the reduced theory can be 'brought back' to the reducing theory, the mental can be 'brought back' to the physical, or heat can be 'brought back' to molecular kinetic energy". In their view, the term 'reduction' as used in philosophy expresses the idea that "if an entity x reduces to an entity y then y is in a sense 'prior to' *x*, is 'more basic than' *x*, is such that *x* 'fully depends upon' it or is 'constituted by it'"; thus, "saying that *x* reduces to *y* typically implies that *x* is 'nothing more than' *y* or 'nothing over and above' *y*" (van Riel and Van Gulick 2014, intro). This is, however, a very general and rough way of characterizing reduction.

But to try to understand scientific and philosophical reduction is indeed no easy task, due to the fact that despite of some agreement concerning a general way of stating reduction, there is a great amount of discussion and confusion around the idea of reducibility in the sciences and its consequences for the neuro-cognitive relation (cf. e.g. Sarkar 1992; Van Gulick 2001; van Riel and Van Gulick 2014; Richardson and Stephan 2009; Silberstein 2002; Walter and Eronen 2011). The idea of reduction has appeared in different forms in philosophy of science and philosophy of mind. Indeed, many accounts of reduction have been being proposed in philosophy of science and philosophy of mind especially since the late 1920s, and there are still in the present many intense debates about which one is the best theory for describing scientific practice and the reduction of cognitive processes.[2]

There are, however, two accounts of reduction that are still largely dominant and very influential in current debates, especially in philosophy of science and philosophy of mind. The first one is the intertheoretical theory of Nagel (1961, Chap. 11).[3] In his view, reduction is "the explanation of a theory or a set of experimental laws established in one area of inquiry, by a theory usually though not invariably formulated for some other domain" (Nagel 1961, p. 338). The set of theories or experimental laws that is reduced can be called "secondary science" and the theory to which the reduction is proposed the "primary science". The central idea is to establish "deductive relations" (or derivations) between the primary science and the secondary science. The deduction, thus, is between theories or laws and it provides a deductive-nomological explanation, where the primary science/theory explains and predicts the phenomena explained and predicted by the secondary science/theory.

In Nagel's view, there are two kinds of reduction: between theories in the same general domain of explanation and between theories in different general domains. The first type is called "homogeneous" reduction, and the second type, "heterogeneous" reduction (Nagel 1961, p. 339). When the theories in play cover the same domains (are homogeneous), the reduction can be achieved relatively easily because both theories are dealing with identical, or very similar, terms. For example, in the domain of physics, Galileo's laws concerning the motion of terrestrial bodies "were eventually absorbed into Newton's mechanics and gravitational theory, which was formulated to cover both terrestrial and celestial motions" (Nagel 1961, p. 339). In spite of the fact that this are two different kinds of body motions studied by two different

---

[2]More systematic discussions about reduction in science lead back to the logical empiricists, with their emphasis on linguistic analysis and their model of reduction as translation (cf. Carnap 1934). This provided the basis (cf. Bickle 2012, p. 89ff) for ontological reductive attempts concerning mind and brain, such as logical behaviorism (e.g. Ryle 1949) and psycho-neural type identity theory (Place 1956; Feigl 1958; Smart 1959).

[3]Nagel's model is the most prominent among the models of intertheoretical reduction. However, another important model built along similar guidelines is, for example, the one of Kemmeny and Oppenheim (1956).

specific areas of physics, there is no need for applying any concept in one area that is missing in the other, i.e. both deal with the same, or very similar, set of concepts. Consequently, the phenomena occurring in the reduced and reductive theory are very similar. This kind of reduction, according to Nagel, despite of occasionally producing revolutions in human knowledge, generates no special puzzles and raises no serious questions about how it is achieved.

However, when the theories in play cover what appears to be different domains (are heterogeneous—as in the case of neuroscience and cognitive science) and use different terms, then the reduction is more difficult to achieve. More precisely, in this kind of reduction, "the secondary science employs in its formulations of laws and theories a number of distinctive descriptive predicates that are not included in the basic theoretical terms or in the associated rules of correspondence of the primary science" (Nagel 1961, p. 342). To realize a proper heterogeneous reduction following this theory of reduction, it is necessary to establish 'bridge laws/principles' that make the reductive links between the terms possible. Nagel does not specify the exact nature of these principles, which could be arguably characterized in terms of conditionals, bi-conditionals or identities. These bridge principles are needed to connect the terms of the primary science with the different terms occurring in the secondary science, so that the deduction can be performed.[4]

The classical example is the "successful reduction of thermodynamics to statistical mechanics in the nineteenth century" (Nagel 1961, p. 337).[5] During a period in the history of science, as Nagel tells us (1961, p. 343), a number of systematic interconnected laws were established dealing with the thermal behavior of bodies; the science dealing with these laws was called thermodynamics. In this science, there were concepts (such as volume, weight, pressure) and laws (such as Hooke's law) employed in mechanics as well, but also different notions, such as temperature, heat and entropy, not employed in mechanics. Because of this, thermodynamics was regarded for a long time as a special discipline, distinct from mechanics and a relatively autonomous physical theory. However, experimental work in the beginning of the nineteenth century found a reductive connection between thermal and mechanical phenomena. Scientists were able to provide a "satisfactory derivation of the Boyle-Charles' law from assumptions statable in terms of fundamental notions of mechanics concerning the molecular constitution of ideal gases. Other thermal

[4]Bridge principles, or bridge laws, are often discussed in the specialized literature. However, this is not Nagel's terminology. These additional assumptions for connecting terms and making the deduction possible are discussed as "postulated linkages" and they can be of different natures (cf. Nagel 1961, p. 354).

[5]Nagel also claims that "certain parts of nineteenth-century chemistry (and perhaps the whole of this science) is reducible to post-1925 physics" (1961, p. 362). However, he does not discuss in detail any example of such a reduction. Moreover, he discusses aspects of a possible reduction of biology to physics and of psychology to physiology in the context of a discussion concerning the "doctrine of emergence" (cf. Nagel 1961, p. 372). No particular example is discussed in detail also in this context, nonetheless.

laws were similarly derived" (Nagel 1961, p. 343). As a result, thermodynamics lost its autonomy and was finally reduced to Newtonian mechanics.[6]

The second account of reduction, that is still influential in present days, predominantly in philosophy of mind, is the functional theory of reduction (cf. Chalmers 1996; Chalmers and Jackson 2001; Kim 1998, 1999, 2005, 2006; Levine 1983, 1993). This account is one important reaction to Nagel's theory and its variations. The clearest proposal is found in the work of Jaegwon Kim. According to this framework, largely influenced by the functionalist position in philosophy of mind, reduction is achieved through essentially two steps. Firstly, the functionalization of the cognitive 'property' by means of identifying the causal role played by the property, i.e. identifying what causes that cognitive process to occur (e.g. environmental stimuli) and what this process causes in turn (e.g. the behavioral outcome). This first step involves only or essentially conceptual work. For example, the property of 'being in pain' could be defined as being in a "state that is typically caused by tissue damage and trauma and that typically causes aversive behavior" (Kim 2006, p. 553). The definition connects pain with physical and behavioral properties. Another example is the reduction of "the gene" to "the DNA molecule": being a gene considered as "the property of being a mechanism" that performs a "certain causal function, namely that of transmitting phenotypic characteristics from parents to offsprings" (Kim 1999, p. 10).[7]

The second step is a matter of empirical scientific research. It consists in finding out through empirical scientific work what the neural physical mechanisms ('realizers') that play such causal/functional role are. In humans, the reduction of pain would be accomplished when "we are able to identify a 'realizer' of pain so conceived, namely a physical state that fits the functional definition, for that population" (Kim 2006, p. 533). Consequently, if neuroscientific research finds out that the activation of a certain group of neurons is typically caused by tissue damage and in turn causes aversive behavior in humans, then we have a neural reduction of pain in humans.[8] In the case of the gene and the DNA molecule, scientific research shows that it is the DNA molecule that fills the causal role of transmitting phenotypic characteristics from parents to offspring and there is a theory specifying precisely how the DNA molecule is able to perform this causal role. In this way, one can claim that "the gene has been reduced to the DNA molecule" (Kim 1999, p. 10).[9]

Contrarily to Nagel's theory, the neural reduction of pain or the molecular reduction of genes do not require any logical derivation of pain from the particular properties of the neural mechanism, or of the gene from the particular properties of the

---

[6]Many authors followed Nagel's lead and proposed improved models of intertheoretical reduction, for instance Schaffner (1967) and Hooker (1981), which were also influential.

[7]According to Bickle (2006, p. 417), however, this idea of genes being composed solely by DNA molecules is not accurate in the light of contemporary molecular biology.

[8]But, of course, the non-reductionist can simply argue that this is just the case if pain is really identical to the functional role ascribed to it, and he can argue that it is not, so there is no identity. Hence, there is no reduction.

[9]Kim himself (1999, p. 10) acknowledges that this is "an idealized, admittedly somewhat simplistic example". However, he does not offer any other in detail (cf. Kim 1998, p. 25, 2005, pp. 101, 102), as Nagel, for example, does.

molecular mechanism, nor logical or conceptual connections between these properties in general. This framework has been influential in philosophy of mind, especially among anti-reductionists with respect to phenomenological consciousness.

4.1.2 The Challenge to Neuro-Cognitive Pluralist Integration

It is as an attempt to improve these relatively more traditional ways of characterizing reduction that the philosopher John Bickle presents his account, which he claims to be entirely based on current neuroscientific work, instead of philosophical descriptive and normative analysis about reduction.[10] Bickle can be regarded as the most prominent representative of contemporary neuroscientific reductionist ambitions. The neo-reductionist argues that Nagel's classic reductionism and its variations are wrong; he is indeed very critical towards theories of reduction like Nagel's one, because, in his view, they are "infected with philosophical assumptions about what the reduction relation is or must be" (Bickle 2012, p. 88). Moreover, he argues that for philosophers of science such as Ernst Nagel or Patricia Churchland the point was to analyse what was in their view uncontroversial examples of reduction from the history of science and construct a theory of reduction that could be applied to all the sciences. Thus, they refer to examples taken from physics, chemistry and biology, while they do not mention specifically examples taken from neuroscience. Bickle argues that although some of these theories assumed and sometimes emphasized the role neuroscience plays for the reduction of cognitive processes, they generally did not present and discuss systematically examples of a real neuroscientific reduction of any cognitive function (Bickle 2012, p. 91). Besides the lack of concrete neuroscientific examples to argue for the reduction of psychology to neuroscience, Bickle points out that it is not even clear whether reduction is an identical or very similar process that happens across different sciences and through the history of the sciences.

Place, for instance, in arguing for the identity of conscious processes and neural processes, discusses examples such as 'lightning' being identical to 'large scale atmospheric electrical discharges' (1956, p. 47). The primary example in Nagel's (1961, p. 342) work, on the other hand, is the reduction of 'temperature' to 'molecular kinetic energy', as we saw. Of course, Nagel's primary concern was not to discuss the issue of cognition and neural processes; he was interested in scientific practices in general. However, among philosophers of mind that accepted his framework, there seem to have been no attempt to apply it to a concrete case of neuroscientific reduction of a cognitive function. Even in Churchland's case, in which some more clear and concrete neuro-cognitive examples are at least discussed with some detail, the point was not to focus on neuroscientific work, because this was seen as controversial from the start (cf. 1986, p. 8). The strategy was rather to find clear examples in the history of science, especially in certain highly prestigious natural sciences, and apply the reductive framework built in this way to cases in neuroscience and cognitive science.

[10]Cf. Bickle (2003, 2006, 2008, 2012, 2015, 2016), Silva and Bickle (2009) and Silva et al. (2014).

In the case of functional reduction, Bickle is even more critical. According to his view, the scientific examples given are based on general knowledge about biological notions such as DNA molecules and genes, or the boiling process of water. As he points out, these are very elementary examples of scientific processes:

> The account of functional reduction, however, takes its inspiration from elementary school textbook understandings of science! [...] These are the scientific examples we use to inform children about 'our scientific world view'. No advocate of functional reduction has ever tried to apply the basic theory to a real example of professional science. No one, for example, has taken a published research report from a major scientific journal and articulated the experiments and results in terms of the hypothesized two stages of functional reduction. There is a reason why no one has tried to do this. Scientific reduction-in-practice simply doesn't proceed in the way this cartoon sketch of an account claims. (Bickle 2012, p. 98)

Thus, he argues that especially in the case of functional reduction the debates are much more related to logical reasoning and metaphysical issues (cf. Bickle 2012, p. 97) and that what one finds in such literature are above all speculations, expectations and hopes towards future achievements or failures in one direction or another.

Contrarily, the neo-reductionist wants to provide a "very different approach toward articulating the nature of reduction" (cf. Bickle 2012, p. 95), which is based on new work coming directly from current neuroscience. Firstly, he contends that even if the traditional theories of reduction are incorrect, one should not give up on the scientific general reductionist project: reductionism should not be seen in a negative manner in science, it has many attractive aspects; science is improved through reductionist disciplines (Bickle 2003, pp. xiv, 1). But, more particularly, if the goal is to understand specifically the nature of scientific mind-brain reductionism, the correct strategy in his view is to look into the experimental practices of a current reductionist field in real neuroscience (Bickle 2012, p. 98). The purpose is to make a kind of meta-science, as he sees it, and not properly standard philosophy of science, much less standard metaphysics. The work is at the start more descriptive and just after it brings a normative component, grounded in the previous descriptions based on the actual empirical research. Thus, the author argues for a general empirical neuroscientific reductive hypothesis, based on what real neuroscientists are currently doing, and argues that philosophers of mind and of cognitive science should be well informed about neuroscientific results. In his view, there is "a growing schism in both philosophy of mind and philosophy of science, between metaphysically minded and normatively prescriptive philosophers versus philosophers willing to countenance scientific practice and results as scientists present them" (2003, p. xiv). Of course, in his view, the second group endorses the correct approach.

Accordingly, a correct approach to neuro-cognitive reduction needs to be supported by real empirical work in current neuroscience. However, for Bickle, "philosophers of neuroscience" are focusing on "the wrong levels of research, theory, and experiment" (2003, p. xiii). That is, cognitive neuroscience. This area, in Bickle's view, is not the best one to look if one wants ultimate explanations of cognition. Therefore, there is here a major difference between Bickle's theory and MTHC, which regards especially cognitive neuroscience as the major field of inquiry for understanding cognition (cf. Chap. 3, Sect. 3.1.7).

In opposition, for Bickle, there is a more important area of current neuroscience, the mainstream of the discipline, which is at the same time ruthlessly reductive in spirit: molecular and cellular neuroscience. As he states:

> [...] higher-level theorists of mind, especially philosophers, should reorient their interests 'down levels' in the neurosciences. Or, short of that, they should realize that the mainstream core of the current science, the part on which all the higher level cognitive neuroscientific investigations ultimately depend, has a 'ruthless' reductionism built directly into its practice. Furthermore, at this cellular/ molecular level, we 'know a lot about how the brain works' and we are increasingly able to manipulate specific behaviors by intervening directly with these cellular processes and intracellular pathways. (Bickle 2003, p. xiii)

In his view, at the molecular and cellular (lower) level of neural activity a lot is already known and it is false that "lower-level neuroscience cannot explain cognition and complex behavior directly" (Bickle 2006, p. 411; cf. 414). Molecular pathways inside individual neural cells can be linked with cognition and these links "are reductions" (Bickle 2008, p. 37). It is thus a clear example of a reductionist field of scientific inquiry recognized by its own practitioners and by scientists working in related fields (Bickle 2008, p. 34). This area of research began in the early 1990s "with the advent of using bioengineered genetic mutations in living, behaving rodents and the use of a variety of behavioral tasks that experimental psychology had developed for investigating specific cognitive phenomena" (Bickle 2008, p. 36).

The central characteristic of the field, thus, is the "application of transgenic techniques from molecular genetics into neuroscience. These features allow experimenters to mutate any cloned gene in living, behaving mammals, and thereby manipulate key proteins in intracellular signaling pathways" (Bickle 2015, p. 305). These techniques increased the scientific capacity of "manipulation specificity and control" in neuroscience, and generally they "enable a clear picture of not only which neurons have been manipulated but also the specific intracellular signaling pathways affected in those neurons" (Bickle 2015, p. 306). In this way, when manipulations of the organisms are successfully made and they produce significant changes in the related behaviors (which can be measured), one can claim that the neural causal-mechanism explain those behaviors, and thereby "the cognitive functions those behaviors are taken to indicate", i.e. "cognitive behaviors" (Bickle 2015, pp. 305, 306). For the neo-reductionist, research in this area, accordingly, "tests, directly and experimentally, causal-mechanistic hypotheses that purport to explain cognition" (Bickle 2015, p. 306).

In Bickle's view, the development and application of these techniques of engineering genetically mutated mammals, together with the development and applications of the relatively recent technique called 'optogenetics',[11] in order to explain cognition in neuroscience (especially in cellular/molecular neurobiology and behavioral neuroscience) can be considered genuine scientific "revolutions" (2016, pp. 1,

[11]Roughly, this technique consists in "using light stimuli to activate or inhibit specific selected neurons in vivo, while the animal engages in behavioral tasks" (Bickle 2016, p. 2).

2). Moreover, within this research fields it is molecular biology, and not physics, the appropriate ideal of genuine science, which should be taken as the primary exemplar of a successful area of science (Bickle 2016, pp. 1, 12, 13). The author further suggests that molecular biology should now replace physics as the "paradigmatic science", against which all others can be "compared and judged" (Bickle 2016, p. 13).

Besides, researchers interested in this field of molecular and cellular neuroscience of cognition publish in high quality scientific journals, such as *Cell*, *Neuron*, the *Journal of Neuroscience*, the *Journal of Neurophysiology*, as well as in the neuroscience sections of *Science* and *Nature* (Bickle 2006, pp. 420, 427). And there is even a professional society for the researchers in the field, the *Molecular and Cellular Cognition Society*. Hence, as the twenty-first century leading neuro-cognitive reductionist points out:

> [...] no one should be mistaken about the *factual existence* of a ruthless and audacious reductionism that informs neuroscience's current cutting edge. If I can communicate that, I will at least break the popular but mistaken myth among philosophers and cognitive scientists that reductionism is 'dead'. On the contrary: it is alive and thriving, at the very core of one of the hottest (and best funded) scientific disciplines. Perhaps I can even convince some that this 'ruthless reductionism' is the *correct* way to pursue a science of mind, given all we know and can do now. (Bickle 2003, p. 5—highlights in the original)

I will call the reductionist theory of human cognition proposed by Bickle *Molecular and Cellular Theory of Human Cognition* (MCTHC). The clearest and most detailed example discussed by Bickle in order to give support for his theory is also related to the memory system, which has provided the field of molecular and cellular neuroscience with its "most impressive achievements" (Bickle 2008, p. 36). To accomplish a reduction in Bickle's sense two important steps are needed. The first one is to intervene using molecular genetic methodology into the genome of animals, usually mice. The aim is "to increase or decrease in vivo gene expression and subsequent protein synthesis of intracellular signaling molecules known to be components of pathways that induce and maintain activity-driven synaptic plasticity" (Bickle 2012, p. 100). These molecules are, for instance, cyclic adenosine monophosphate (cAMP), response-element binding protein (CREB) and protein kinase A (PKA). There are different ways of intervening to modify the expression of these molecules: (1) intervention to decrease molecule expression; i.e. the genes for such specific molecules can be made non-expressive (knocked out) either permanently during the embryonic stem cell stage of development; or temporarily in the adult animal; or (2) intervention to increase molecule expression; i.e. additional copies of the genes can be inserted using transgenic techniques to increase the amount of protein in vivo (Bickle 2012, p. 100). Techniques of control can be employed to make sure that "the genetic manipulation and subsequent protein synthesis were induced correctly and to isolate the specific cognitive function and control for confounding cognitive effects" (Bickle 2012, p. 101).

The second step is to measure the effects of the intervention in the behavior of the organism under controlled experimental conditions. The genetic modified animals perform a variety of behavioral tasks so that their specific 'cognitive functions', linked with the specific kind of behavior, can be measured. Their behaviors are contrasted with the control animals who are not genetically manipulated. The significant behavioral differences are then understood as the result of the genetic manipulations: the differences in the genetic mechanisms of protein production is the most relevant direct causal factor for explaining the modification in the cognitive function that ultimately produces the behavioral differences (Bickle 2012, p. 100).

According to Bickle (2012, p. 101), some scientific experiments already present evidence for establishing the connection between a molecular and cellular mechanism and a particular behavior that indicates a cognitive function. In one scientific experiment a mouse in which CREB was 'knocked-out' had intact short-term memory on many rodent memory tasks, while in the long-term memory versions of these tasks there were a great decrease in the memory capacity in comparison with the control. In another experiment CREB was increased in a small population of neurons in a manipulated mouse. This led to an increase of memory consolidation measure by fear conditioning behavior in the modified mouse. Since CREB has been traditionally considered to be implicated in the induction of long-term potentiation (LTP—a form of neural activity that can last hours, days or weeks which increases neurotransmission efficacy at individual chemical synapses; cf. Chap. 3, Sect. 3.2.2), ultimately is the presence of CREB that is doing the bottom-level most central causal work. In fact, CREB is part of a particular molecular mechanism that involves cAMP, PKA and CREB, which leads to LTP. Bickle claims that blocking any step of this mechanistic process "virtually eradicates memory consolidation, while enhancing steps can lead to faster and stronger consolidation" (2008, p. 38).

The central idea is that these dynamics at the molecular level is actually what *produces* the cognitive processes normally called long-term memory and memory consolidation. What is being asserted is that there is already sufficient empirical experimental evidence to establish a "causal connection between a proposed cellular or molecular mechanism and a complex, system-level cognitive phenomenon" (Bickle 2012, p. 102; cf. Silva and Bickle 2009). These causal connections are established through experiments that: (1) manipulate the animal in order to increase the probability, duration or extent of the molecular mechanism and measure the probability, extent and duration of the behavioral effect; (2) manipulate the animal in order to decrease or eliminate the molecular mechanism and measure the effects; (3) do not manipulate, but seek to find correlations between the molecular mechanism and the behavior; (4) analyze the empirical evidence available integrating the results (including those from 1, 2 and 3 when available) and seeking to establish causal relations between the molecular mechanisms and the behavior.

In order to establish these causal connections, though, Bickle admits that "higher-level scientific investigations" are necessary (2012, p. 103). This is because precise knowledge about how the whole system behaves is important in order to correlate the "proposed molecular mechanism and the system's behavior we use to indicate the occurrence of a specific cognitive function" (Bickle 2012, p. 103). Behavioral experiments at a higher-level also help to establish the "theoretical plausibility of the proposed molecular mechanism for that cognitive phenomenon" (Bickle 2012, pp. 103–104). These are jobs for cognitive scientists and animal experimental psychologists according to Bickle. Moreover, cognitive neuroscience and its mechanistic goals of decomposition and localization is also important, since it is necessary to identify the most relevant types of neural activity. Consequently, as one can see, there is plenty of job for cognitive experimental psychologists and cognitive neuroscientists, which usually inquire about cognition working at higher-levels. In this respect, then, the 'ruthlessly' component of Bickle's reductive approach becomes instead much more 'friendly'.

Nevertheless, his reductive approach stands on one very important point: in spite of the necessity of higher-level scientific inquiry, the "hypothesized molecular mechanism is actually doing the causal work", i.e. "the best causal-mechanistic story for the specific cognitive function then resides at the lowest level of effective experimental interventions" (Bickle 2012, p. 104). Not all levels are equally significant for the explanation, there is no pluralist integration, but indeed a reductive one. As he states: "The *explanatory* relevance of intervening levels is no longer needed when the 'intervene cellularly/molecularly and track behaviorally' approach succeeds" (Bickle 2006, p. 426—highlight in the original). In this account, higher-level fields such as cognitive neuroscience and cognitive science would be ultimately merely descriptive endeavors generating data, but not genuinely explanatory (cf. Bickle 2006, pp. 427–428). This means that there is no explanatory pluralism here, and this is in sharp contrast with the mechanistic pluralist theory. For Bickle, what counts in the end for providing the scientific explanation is the molecular and cellular level:

> [...] one way in which their view [the view of the neo-mechanists] differs from the reductionism espoused here is their emphasis on *multi-level* mechanisms. The question is whether multi-level mechanisms are still recognized *as mechanisms* when neuroscience has successfully 'intervened cellularly/molecularly and tracked behaviorally'. I contend that they are not [...]. (Bickle 2006, p. 430—highlights in the original)
>
> [...] I contend that my 'intervene cellularly/molecularly and track behaviorally' account of reduction better captures the practices of cellular and molecular neuroscience than does the analysis of the new mechanists. (Bickle 2006, p. 431)

Thus, in Bickle's view there are no multi-level mechanisms in ultimate scientific explanations of cognition (including human cognition) in neuroscience; there are solely molecular and cellular mechanisms of cognition.

In a nutshell, Bickle argues against intertheoretical accounts of reduction that emphasize deduction between laws concerning theories from different fields—on this point there is agreement between Bickle and the multilevel mechanists. He also argues against reductive theories that emphasize the functionalization of concepts and their physical structural implementation, without providing detailed genuine examples of this kind of reductionist process in actual science. He argues instead for a neuro-cognitive reductive approach based on experimental work currently done in neuroscience, i.e. neuro-cognitive reductive scientific integration instead of neuro-cognitive pluralist scientific integration, contrarily to the mechanistic theory. His favoured field of research is molecular and cellular neuroscience, which in his view is entirely reductionist. The more important example discussed by the author concerns the memory system in rodents. The example given aims to show that there is a direct causal connection between a particular molecular mechanism and a specific cognitive function responsible for a specific kind of behavior. Many scholars consider this view as the clearest and strongest theory of neuro-cognitive reduction currently available, especially because it is very close to actual contemporary neuroscience. It is true that Bickle's framework focuses its support on examples of cognitive processes generated by non-human animals, especially mice. However, his intention is to use such examples to draw broader conclusions concerning the nature of general natural cognition, including human cognition. To this extent, it is correct to regard the framework as presenting an influential contemporary theory about the nature of human cognition and its explanation.

4.1.3 The Mechanists' Reply: Standing with Pluralism

The relationship between the mechanistic theory and neuro-cognitive reduction is highly ambiguous. On the one hand, the theory is clearly committed to pluralism: its proponents argue for multilevel integration in order to achieve unity in science (cf. Chap. 3, Sect. 3.1.7). Accordingly, the mechanistic theory forcefully rejects the classical theory of intertheoretical reduction proposed by Nagel (1961). Bechtel and Craver argue that the classical theory, as well as its subsequent versions, cannot be successfully applied to the biological sciences. This is the case due to the commitments of Nagel's theory to deduction and general natural laws, given also that his theory of reduction is based on DNTSE, favored by the logical empiricists of the last century. Contrarily, the mechanistic theory does not focus on logical deduction and laws, which, as they argue, are typically not useful for the concrete work of neuroscientists; in contrast, the mechanistic theory describes mechanisms (cf. Chap. 2, Sect. 2.1.3).

However, on the other hand, one of the most important proponents of the mechanistic theory assumes a kind of 'reductionist' stance. Bechtel claims that "from the point of view of mental activity" his approach is reductionist, and he calls it "*mechanistic reduction*" (2009, pp. 13–14—highlighted in the original). As he clearly states: "Mechanistic explanation, in seeking to explain the behavior of a mechanism in terms of the operations of its parts, is committed to a form of reduction" (Bechtel 2008, p. 129). According to this view, "going down a level offers a kind of reduction (to component parts and operations)" (Bechtel and Abrahamsen 2005, p. 426), i.e. the "clear sense" in which "mechanistic explanations" are "reductionist" is to the extent that they focus on "decomposing a mechanism into parts and operations that explain why the mechanism behaves in a particular way" (Bechtel 2008, p. 142; cf. Bechtel and Wright 2009, p. 126).[12]

[12]Bechtel's mechanistic theory is also reductionist from the ontological point of view, due to his commitment to mind-brain type-type identity theory (cf. Chap. 3, Sect. 3.1.3). Each time a genuine cognitive function is 'localized' in the brain as a function of a physiochemical neural mechanism (i.e. identified with the function of a particular neural mechanism), this is an ontological reduction as classically understood in philosophy of mind. However, the discussion here concerns primarily the reduction of cognition to different levels of neural explanations (epistemological explanatory

In Bechtel's view, the mechanistic reduction "occupies a middle ground between vitalism", or any other kind of dualism, which "sees no hope of reduction", and "the exclusionary account of reduction, which views all explanatory work as performed at the lowest level (a position that Bickle 2003, calls 'ruthlessly reductive' and vigorously defends)" (Bechtel 2008, p. 130). How exactly, then, reduction should be understood in Bechtel's view? Bechtel and Richardson (1992) give the answer. In their view, a reductionist working in psychology wants to explain phenomena at one level of organization, that belongs to one field of science, in terms of lower-level mechanisms, that belong to another field of science (Bechtel and Richardson 1992, p. 261). The notion of levels, for the authors, is understood in terms of part-whole; thus, a lower-level discipline would focus on individual components of a given system, while the higher-level discipline focus on the whole system. On this account, a neuroscientist trying to explain psychological processes by using concepts and tools from neuroscience would count as a reductionist (Bechtel and Richardson 1992, p. 261).

One can note, therefore, that the issue concerning reduction in the life sciences is to determine at what level to perform investigations in order to explain the phenomenon under consideration with its most important and complete physical causes. Thus, the idea is to focus on properties of the components of mechanisms and the causal processes and functions related to these properties that are necessary for the ultimate explanation. As Bechtel himself indicates, reduction in the life sciences is a matter of 'lower-level' components, activities and organization explaining 'higher-level' components, activities and organizations. In this sense, MTHC attempts to be *weakly reductive*, since it argues at the same time for plurality of levels to construct weakly autonomous scientific explanations in the life sciences. Thus, the *strong neuro-cognitive reductionist framework* defended by Bickle appears as a "competitor" to "multi-field integration in neuroscience" (Craver 2005, p. 393).

For the multilevel mechanists, the real problem with 'strong' (Bickle's) neuro-cognitive reductionism is due to the fact that "whole systems exhibit behaviors that go beyond the behaviors of their parts" (Bechtel 2008, p. 129). To put it in another way: "It is the complexity of the organization that permits the lower-level components to produce higher-level behavior, so we cannot simply appeal to lower level parts to explain the system's performance" (Bechtel and Richardson 1992, p. 285, cf. p. 279). In another passage, Bickle's approach is characterized as follows: "The cellular and molecular processes targeted by the ruthless reductionist are typical not operating parts of these mechanisms, but are parts of their operating parts (or of the operating parts of the operating parts, etc.)" (Bechtel 2009, p. 14; cf. p. 21). For this reason, they are at a 'lower level' of the mechanism's organization.

One can clearly see, therefore, that the strong reductive position is being characterized as defending the thesis that 'individual component or subcomponent parts can (always or frequently) explain alone the behavior of a given whole'. Thus, Bickle risks "ignoring the other components of the mechanism and the organization that

reduction). I will discuss further the issue of ontological reduction through identities in Chap. 6, Sect. 6.1.3, B.

enables the components to work together to produce the phenomenon of interest" (Bechtel 2009, p. 21). In this regard, Bechtel continues, it is necessary "not just to identify a single component but to identify many of the parts and operations so that it is possible to conceptualize how the operations couple together to realize the phenomenon" (2009, p. 22). Bickle's reductionism, thus, fails according to the mechanists because "it focuses on only one or a few subcomponents within the mechanism and fails to consider how those subcomponents are related to others in the realization of the phenomenon in question" (Bechtel 2009, p. 22).

Instead, Bechtel wants to be a 'weak reductionist'—contrarily to the extreme reductionist position that he characterizes, i.e. Bechtel's version of Bickle's reductionism. This weak reductionism can also be understood as a kind of 'weak emergentism' (i.e. not a 'dualist' emergentism).[13] For Bechtel, it is the specific organization of a given system that permits it to perform functions that its components alone are not able to perform. Since the components of systems interact, one cannot simply focus on one component in order to explain the function of the whole. Craver's ideas on this point are essentially identical. In his view, "mechanisms can do things that individual parts cannot", "mechanisms explain things that individual parts cannot", thus "higher levels of mechanisms are legitimately included in the explanations of contemporary neuroscience" (Craver 2007, p. 216). As he points out:

> I do not advocate a spooky form of emergence. It is important to keep several different senses of the term 'emergence' distinct. Some philosophers and scientists use the term 'emergence' to describe properties of wholes that are not simple sums of the properties of components. Mechanisms are non-aggregates, and so they are emergent in this weak sense. Mechanisms require the organization of components in cooperative and inhibitory interactions that allow mechanisms to do things that the parts themselves cannot do. (Craver 2007, p. 216)

In this sense, "individual lower-level components do not explain the overall performance of the mechanism. Only the mechanism as a whole is capable of generating the phenomenon" (Bechtel 2008, p. 146). Accordingly, Bechtel and Craver intend to claim that identifying components, operations and organization is important to explain the phenomena produced by the whole mechanism. Consequently, "the behavior of the whole system must be studied at its own level with appropriate tools for that level", since this level has "a kind of independence" and the phenomena are "different from those studied at the level of the component parts" (Bechtel 2008, p. 129). Therefore, the first line of argument is the emphasis on the importance of the internal complete organization of a given whole, which in their view already undermines strong reduction:

> Thinking of higher levels as organized helps to highlight the crucial role organization plays in making functioning mechanisms out of parts, and to emphasize that an adequate scientific account of a mechanism requires more than identifying parts and operations at a single lower level. (Bechtel 2008, p. 130)

Accordingly, in their view, even Bickle's most compelling example related to the memory system does not consist of a top-down reductive search for lower-level

[13] In fact, Bechtel's sympathy towards emergence is old (cf. Bechtel and Richardson 1992, p. 285).

mechanisms of memory; this picture, according to multilevel mechanists, is simply inaccurate and misleading. The classic memory example can rather be understood in accordance with the pluralist view of cognitive neuroscience integration (Craver 2007, p. 237ff). In this view, the LTP neuro-physiological process is considered to be part of the mechanism that produces (and explains) memory consolidation, not as identical to the whole mechanism that produces memory consolidation, nor as a kind of memory consolidation. As they argue, LTP is important for memory consolidation because it can be induced in the hippocampus, which is a crucial brain region for various sorts of memory capacities, as shown by PET, fMRI and other methods. Besides, data from lesioned humans and nonhuman animals show that, when the hippocampus is damaged, there is a great decrease in memory consolidation capacities (consider for instance the famous case of H.M., and experiments done with rats in mazes when performing memory tasks).

This is regarded as compelling evidence for the claim that the hippocampus is central for the memory mechanism, at least it is a central part of a mechanism that can be even more large and complex (cf. Craver 2007, pp. 260–261). Since the hippocampus is strongly correlated to memory consolidation, a mechanistic description of its components and connections is necessary. Moreover, it is very likely that there will be other important causal processes to be discovered. LTP is then considered to be a component in a larger mechanism that produces a behavior that is thus identified with memory consolidation. Accordingly, the particular molecular mechanism that involves cAMP, PKA and CREB, which leads to LTP is thus a tiny component of a larger memory mechanism for memory consolidation and cannot alone be considered the ultimate causal explanation for the phenomenon, as Bickle argues. Consequently, there is at least one important higher-level, i.e. other processes happening in the hippocampus, being left out in Bickle's reductive explanation. It is in this sense that the higher-level provides new information about the mechanism (cf. Bechtel 2007, p. 174); information that cannot be found in the lower level, since the lower level is portrayed as dealing partially with just component parts.

Another dimension of the debate concerning the internal organization of a given whole is related to predictability, and arises when we consider wholes which are highly complex systems. There are systems that are not so complex, as for example, a bike, or a car, or even an airplane. If one wants to understand fully the functions and behaviors of such systems, all one needs to do is to look to their component parts and subparts to understand what they do and how they interact causally with each other. This knowledge about all the components of the system can fully explain everything one needs to know about the system, and one can predict its behavior with a high degree of accuracy in this way, as long as appropriate knowledge of external conditions is also provided. To illustrate this, we can think about an airplane under turbulence, or a car being driven in a wet road due to a raining day. In such cases, explanations concerning how these simple systems will behave can be very accurate, as well as the related predictions.

In highly dynamical complex systems, there is no linearity in the causal interactions and the components of the system relate not in a static, but in a dynamical way, which makes the system to change constantly its own state and the way it is related to the environment. In such systems accuracy in explanations and predictions is not so high. The complex interactions produce new phenomena that are radically different from those related to the components of the system and which cannot be fully predicted or explained just by the knowledge of the operations associated with those components taken in isolation. This occurs because the variables are too many and they cannot be measured or understood fully. There will be always a degree of imprecision, a degree of uncertainty in the final value of measurement. And since the initial conditions of a system cannot be measured with complete accuracy, i.e. the initial measurements will always have a degree of uncertainty, the results derived from those measurements will also be uncertain. Indeed, if the system is too complex, the measurements of their parts will have a larger degree of inaccuracy, and the longer the time lapse, the more those inaccuracies will have an effect on the final results, leading with time to a chaotic state of indeterminacy.

In this case, the future behavior of the system can hardly be predicted, as occurs with weather, for instance, or with the dynamical evolution of large ecosystems. The more complex the system is in terms of variables and complex causal interactions and the more time passes, the more inaccurate the results will be in terms of expected predictions. In the case of the issues related to brain and cognition, if the brain (or parts of the brain) could be considered as such a complex dynamical system (as some authors argue),[14] it would be impossible to predict mental phenomena arising from it just by looking at physio-chemical neural operations and interactions of the components of this brain, or brain region. This is because it would be impossible to know all the initial conditions and to measure all the many variables in the brain that are responsible for the generation of a certain mental phenomena with total accuracy, since it would be so complex. This inaccuracy in the measurement always leads to inaccurate results in any attempt to deduce or derive mental phenomena from neural activity. Consequently, any completely accurate prediction of complex human behavior related to mental phenomena would turn out to be impossible. Some authors argue that reduction is incompatible with such unpredictability, because if one does not have all the relevant knowledge of the elements of a system that 'behaves' in a certain way, one cannot anticipate (predict) all the aspects of this behavior. Thus, the idea is roughly: no accurate explanation and prediction, no reduction. This is then another reason why reduction fails: certain mechanisms/systems (wholes) with highly complex and dynamical internal organization behave in a way that cannot be predicted with high accuracy. As Craver points out:

> Some mechanisms have so many parts and such reticulate organization that our limited cognitive and computational powers prevent us from making [...] predictions. Some mechanisms are so sensitive to undetectable variations in input or background conditions that their behavior is unpredictable in practice. Behaviors of mechanisms are sometimes emergent in this epistemic sense. (Craver 2007, pp. 216–217)

Similarly, Bechtel and Richardson (1992, p. 266) claim that the behavior of some systems can be novel and unpredictable considering the behavior of the parts in isolation, since the particular organization of the system gives rise to particular properties;

[14]For instance, Gazzaniga (2012).

in these cases, these properties are "emergent", because they are unpredictable, and this happens generally with nonlinear systems, which have a more complex mode of organization.

Moreover, there is yet another dimension of this debate on reduction. As Bechtel and Wright emphasize, "a given mechanistic activity is always constrained by its environmental conditions" (2009, p. 127; cf. Bechtel 2008, p. 148). To put it in another way, a particular whole mechanism behaves in certain way "only under appropriate conditions" (Bechtel 2008, p. 146). Therefore, the context in which the mechanism behaves is another important aspect for understanding its behavior, not just the internal components, operations and their organization (cf. Bechtel and Abrahamsen 2005, p. 426). It is thus crucial for a correct understanding of the behavior of a mechanism to identify the external factors that can vary and affect it. These factors can be in turn understood as components in a larger mechanism, in which the target mechanism is embedded. It is this mechanistic environment that provides conditions for the rise of the particular behavior. Often, thus, an explanation must clarify what the appropriate environmental conditions for the appearance of a given phenomenon are:

> No matter how much they investigate the parts, their operations, and their organization, investigators will not identify the variables in the environment that are impinging on the mechanism. Discovering these variables and their effects requires inquiry directed at the environmental variables using appropriate investigatory techniques. (Bechtel 2008, p. 152)

In order to do this, enquiries concerning the higher-level of the whole mechanism is crucial, as Bechtel points out:

> [...] the recognition that parts and operations must be organized into an appropriate whole provides for a robust sense of a higher level of organization. A scientist seeking an account at this higher level will find it essential to undertake independent study of the organization of the mechanism and how it engages its environment. (Bechtel 2008, p. 130)

Another line of attack against reduction is to argue that even some cases of reduction in general science are controversial as cases of reduction. And even granting that scientific reduction in the described way at least happens, the cases are rather rare and peripheral to science (Craver 2005, p. 377). It has been also argued that so far there are no clear cases of reduction of any psychological theory to a neuroscientific theory about a given mechanism. These arguments, however, are not going to convince Bickle. He could simply answer that his theory of reduction concerning the capacity of memory provides a clear case where reduction occurs in the neuro-cognitive sciences. He would also argue that reduction here is not peripheral and that it is on the contrary ubiquitous in neuro-cognitive science. Therefore, those arguments cannot be used with so much success in order to undermine Bickle's reductionism.

Finally, the neo-mechanists point out that molecular and cellular neuroscience developed recently new methodologies, which include techniques for the manipulation of structures and processes at the low level of genes expression, molecules and proteins. According to them, this is the most important reason why this area is increasingly growing, not because neuroscientists agree that molecular and cellular levels are the most important for the causal explanation of cognitive processes. As Craver points out:

> I do not dispute that molecular biologists are making very exciting contributions to contemporary neuroscience. Nor do I dispute that the proportion of neuroscience dedicated to molecular pursuits has expanded dramatically in recent years. Even if one grants that there is an historical trend toward the molecular, there is a further question of what hypothesis best explains that trend. Bickle's hypothesis is that only molecular explanations are truly explanatory, and that neuroscientists are 'going molecular' because that is where the true explanations are. However, an alternative hypothesis is at least equally plausible. Researchers have recently developed a host of new techniques for sequencing, copying, and manipulating genes, and for designing pharmaceuticals that control molecular mechanisms. These techniques allow molecular biologists to answer myriad questions about the molecular constituents of nerve cells that could not have been posed even a few years ago. As a result, the field of molecular biology has expanded greatly. (Craver 2007, p. 235)

Thus, there is a clear methodological dimension here, since the availability of new and better methods for a science can certainly make a great contribution for its growth. Cognitive neuroscience also had a great growing with the development of PET and fMRI imaging techniques. When new methodologies come out, there is excitement, funding and growing (this is a salient trait in the sociology of the sciences).

However, if this point is used against Bickle's reductionism as an attempt to refute it, it will not succeed, because it is irrelevant for that purpose. To argue that molecular neuroscience current popularity is due to methodological advancements is beside the point. That can well be true, and probably it is. However, it simply does not entail that Bickle's neuroscientific reductionism is wrong. We are left in the end with three most important and plausible arguments against Bickle's reductionism: (1) related with the internal organization; (2) related with unpredictability of complex systems; and (3) related with external factors that affect the behavior of the mechanism. With these arguments, the mechanists attempt to defend plurality as a form of integrating and unifying neuro-cognitive sciences, at the same time they attack reductionism.

To sum up, the mechanistic theory argues that reductive approaches focus exclusively on explanations that appeal to lower-level components of whole mechanisms, and that this is misleading, since adequate explanations in sciences interested in cognition typically include higher-level causes, i.e. organization, especially in complex systems, and external factors which also often makes cognitive mechanisms to be complex and their behavior emergent in a 'non-spooky' way. For them, reductive theories are, therefore, inadequate to account for the integration of neural and cognitive science, because progress in the life sciences occurs rather by integrating theories 'pluralistically' rather than 'reductively'. Different neuro-cognitive fields constrain a multilevel mechanistic explanation, providing thus the necessary tools for a plural integration. These fields "are autonomous in that they have different central problems, use different techniques, have different theoretical vocabularies, and make different background assumptions" (Craver 2007, p. 231). Accordingly, the neo-mechanistic authors insist with the pluralism concerning cognition and the brain: "causal processes at different levels in nature are generally quite different in character and one must develop appropriate vocabularies to describe the particular causal interactions at any given level" (Bechtel 1994, p. 19). Ultimately, thus, the theory stands for pluralist levels of explanation and pluralist levels of causation, not for any kind of genuine explanatory reduction.[15]

4.1.4 Emergent Phenomena and the Mechanistic Theory

Is this answer satisfying? Let us consider it more carefully. First of all, what exactly is the idea of reduction the theory is attacking? Is it correctly characterized? What exactly is the notion of weak emergence that the theory is defending? Is it really incompatible with a robust notion of reduction, such as Bickle's one? The advocates of the mechanistic theory argue for some quite simple ideas: that the behavior of a whole is more than the behavior of its parts; that the organization between all the components needs to be considered; that unpredictability undermines reduction; and that environmental external relations/factors influence the behavior of the whole. However, such ideas and other aspects of the relationship between the whole and its parts have been subject of intense debates already for a long time, especially in the literature on emergentism, which is a topic being deeply discussed in philosophy of mind and philosophy of science since even before the second half of the twentieth century. Many emergentists frequently appeal to the compositional relations between whole and parts to make their cases.

The concept of 'emergence' is highly controversial; therefore, it is extremely important to start with a clear analysis of what it really means. The term comes from the Latin verb *emergo*, which means 'to arise', 'to rise up', 'to come up' or 'to come forth'. Since the second half of the 19th century, most prominently the British philosophers John S. Mill (1806–1873), Alexander Bain (1818–1903), George H. Lewes (1817–1878), Samuel Alexander (1859–1938), Lloyd Morgan (1852–1936) and C. D. Broad (1887–1971) have made emergence the core of a comprehensive philosophical position (for an overview cf. McLaughlin 1992).[16] For these authors, physical phenomena in nature and the related sciences responsible for investigating them are organized according to different hierarchical levels of complexity. At the bottom level there is physics and at higher-levels there is chemistry, biology, psychology and sociology. One of the main points of debate in which these scholars defended emergentism was related to chemical phenomena and their possible reduction to physical phenomena. It was argued by the emergentists that in a given chemical reaction such as $CH_4 + 2O_2 \rightarrow CO_2 + 2H_2O$ (methane plus oxygen produces carbon dioxide plus water) the properties of the chemical components being produced could not be completely explained or deduced given the properties of the chemical components producing it (McLaughlin 1992, p. 60).

[15]Craver indeed rejects completely any epistemological dimension of mind-brain reductionism (2007, p. 228ff.).

[16]The idea of emergence spread through many countries in this period. Interestingly, McLaughlin counts William James among U.S. emergentists (1992, p. 57). However, he does not examine any of James' views on the topic.

Another important point of debate was related to the phenomenon of 'life' and the relationship between biology and chemistry (cf. Chap. 2, Sect. 2.1.2). The issue they addressed was whether 'life' should be understood in a simpler physio-chemical way and whether biology could be reduced to 'lower level' chemistry and finally to physics. The emergentist position developed in this period was intended to propose a middle way (cf. Stephan 1992, p. 26), eliminating a certain vital immaterial substance usually called 'entelechy' and, at the same time, retaining, in some sense, irreducibly vital processes. The internal structures of the organism are constituted only by physio-chemical processes; however, since the organization of these processes is so complex, emergent vital processes arise that are in a higher-level than the physio-chemical constituents. According to the emergentist position, life would be a genuinely new property of the organism as a whole that arises (emerges) from the interaction between the organism's parts, but which could not simply be reduced to the parts and the relations holding between them. Note that this is exactly where Bechtel and Craver want to stand with their mechanistic pluralist position, i.e. in the middle between Bickle's ruthless reductionism and dualist views that bring souls or 'spooky stuff' (as they put it) to the picture.

Finally, also the relation between physiology and psychology was addressed. The emergentists, such as Samuel Alexander and C. D. Broad argued that particular mental phenomena emerge from a certain configuration of physiological phenomena, but are somehow not reducible to this physiological base from which they have emerged (cf. Stephan 1992, p. 31). The "examples par excellence are the good old secondary qualities", such as color, taste, smell, and sound (Stephan 1992, p. 39).

In contemporary debates, the term 'emergence' is used in a variety of technical senses, and there is confusion about what is really meant by 'emergence' or by an 'emergent property'. A vast specialized literature has been written in order to debate and clarify the concept (cf. e.g. Beckermann et al. 1992; Bedau and Humphreys 2008; Corradini and O'Connor 2010; Crane 2001; O'Connor 1994; O'Connor and Wong 2015; Macdonald and Macdonald 2010; Stephan 1997, 1999, 2002). Roughly, however, the basic idea of emergence is that the features of a complex system emerge from the features of its component parts taken together. In this sense, the features of the whole system are more than the features of its components parts taken in isolation; they are not completely independent of those parts, since they emerge from them, but they are over and above them (cf. Van Gulick 2001).

Emergent properties 'arise' out of more fundamental properties and yet are novel or irreducible with respect to them. A property is emergent when it arises from a system that has reached a certain level of complexity. The idea is that a dynamical interaction in a highly complex system can lead to the emergence of particular properties. Emergent properties exist only insofar as the fundamental system exists. At the same time, emergent properties are distinct from the properties of the parts of the fundamental system from which they emerged. Therefore, emergent properties are systemic properties, i.e., properties that are exhibited by the system but that are not exhibited by any of its parts considered individually or in groups that are smaller than the whole system (e.g. being alive is a property that characterizes a dog as a whole organism, but not its parts taken individually or in groups smaller than the whole system—at least none of its subcellular parts; cf. Van Gulick 2001, p. 17).[17] Besides, systemic properties are also more than the mere 'sum' of the individual component properties (e.g. by contrast, the mathematical sum of the weights of the individual parts of a car is equal to the weight of the whole car; therefore, this is not an emergent property).

Frequently, two main types of emergence are distinguished in the literature: weak and strong (ontological) emergence.[18] In Chalmers' view, one can say that a "high-level phenomenon is *strongly emergent* with respect to a low-level domain when the high-level phenomenon arises [...] from the low-level domain, but truths concerning that phenomenon are not *deducible* even in principle from truths in the low-level domain" (2006, p. 244). According to the author, this is the notion "most common in philosophical discussions of emergence" and it is the same notion "invoked by the British emergentists of the 1920s" (Chalmers 2006, p. 244). This phenomenon, or property, cannot be deduced from the lower level, since it is completely 'new' and 'fundamental'. This strong emergent mental phenomenon is, thus, supposed to be *irreducible* to the low-level physical domain even though it arises from it (cf. Stephan 1999, 2006, 2013).[19] This requirement for irreducibility is so specific that the only kind of property that is most clearly considered to be strongly emergent in the literature in philosophy of mind is the kind related to phenomenal consciousness. The relation between the lower-level and the higher-level here involves a sort of 'mereological supervenience' (cf. Kim 2006, p. 549), where the components at a lower level and their organization altogether determine the properties at a higher-level. The properties at a higher-level, in turn, are said to be emergent from the physical lower-level but from a different nature. Moreover, most commonly they are said to have causal powers on their own, i.e. to be autonomous with respect to the causal powers of the lower-level components and their organization—otherwise these properties could be considered merely epiphenomenal or explanatorily irrelevant.

The idea of strong emergence has gained ground also among scientists.[20] Indeed, even current leading cognitive neuroscientists, such as Michael Gazzaniga, defend

---

[17]The property of 'being alive' mentioned by Van Gulick can be considered a classical example of emergent property in this context. It can be claimed, though, that the dog is totally composed by cells and each of these cells is alive, or at least the absolute majority. In this sense, the parts of the biological system have the same property as the whole. However, the way in which a dog is considered to be alive is different from the way a cell is considered to be. The dog, for instance, is capable to desire food, to feel and avoid pain and to engage in a variety of behaviours that its particular condition of being alive permits. In this sense, the sum of 'liveness' of the cells that constitute this biological system will never be equal to the condition of being alive of the whole dog.

[18]Here we are interested in the synchronic kind of strong ontological emergence, i.e. the one that happens in a particular moment of time, and not across time.

[19]It is important to state clearly that emergentists attempt traditionally to be physicalists. Although they argue for the emergence of certain properties and in certain cases their irreducibility to the lower level, these properties are considered physical (cf. McLaughlin 1992, pp. 49, 55).

[20]Forms of emergentism have been defended by some scientists; e.g. the neurobiologist and neuropsychologist Nobel laureate Roger Sperry.

the thesis. Let us consider in more detail this defense, since it is a very interesting example of how the position has been being proposed; thus, it can provide some important clarifications. Gazzaniga is currently one of the most distinguished researchers in the field of cognitive neuroscience. Some authors even consider him to be the 'father' of cognitive neuroscience (Reuter-Lorenz et al. 2010). Thanks to his numerous contributions to the field over a span of more than fifty years, he has had an enormous influence on many significant debates (cf. Gazzaniga 1970, 1985, 1988, 1992, 2000, 2005, 2008, 2012, 2015; Gazzaniga and LeDoux 1978). The importance of Gazzaniga's impact is also due to the fact that he is one of the few contemporary cognitive neuroscientists who tries to explicitly and directly address the debate on the so-called mind-body problem, a problem which tacitly underlies all discussion in cognitive neuroscience since this discipline ultimately interrogates the relationship between the neural system and cognition. In his view, "the foremost objective of the brain sciences is, of course, to determine the relation between mind and brain" (Gazzaniga and Ledoux 1978, p. vii). Given Gazzaniga's stature in the field of cognitive neuroscience it is very important to consider such approach.

He believes that "the brain enables the mind", i.e. our mental life reflects the actions of neural devices in our brains (Gazzaniga 2000, pp. xii, xiii; cf. 2005, p. xiv; cf. 2012, 2015). However, even though Gazzaniga agrees with the idea that we are nothing but biological machines living in a physical world (Gazzaniga 2012, p. 7), he does not consider necessary to subscribe to a reductionist account. Instead, he attempts to defend a physicalist but at the same time nonreductionist position. In his view, strong emergentism is the best account so far for the relationship between cognitive and neural phenomena. Emergence, as he understands it, is when micro-level complex systems self-organize into new structures, with new properties that previously did not exist, to form a new level of organization on the macro-level (Gazzaniga 2012, p. 124; cf. 2015, p. 343). Moreover, Gazzaniga (2012, p. 124) believes that there are two schools of thought on emergence, a weak one and a strong one. According to his view, in weak emergence, new properties arise due to interactions in the system and the emergent property is reducible; however, in strong emergence the new property is irreducible, since it is more than the sum of parts and since there is unpredictability in the system from the individual elements to the whole (Gazzaniga 2012, p. 124).

In Gazzaniga's view, neuroscience alone is not able to explain mental phenomena as they are explained by psychology or cognitive science, since mental phenomena are emergent phenomena (which arise from our underlying neuronal, cell-to-cell interactions, but cannot be predicted and understood by knowing only about cellular interactions). Thus, in his view, psycho-neural reduction will never succeed (Gazzaniga 2012, pp. 107, 130). The link between the emergent phenomena and their bases is, for him, not to be understood in causal terms. There is no upward nor downward causation here. The link must but just accepted as a brute fact, some sort of unanalyzable connection, where no further explanation needs to be put forward. As Gazzaniga states: "[…] the brain enables the mind in some unknown way"; "I think conscious thought is an emergent property. That doesn't explain it; it simply recognizes its reality or level of abstraction" (2012, p. 129, 130).

Indeed, strong emergence can be considered an attractive position in many ways. Emergence seems to have "a special appeal for many people"; it is not like reduction, which "sounds constrictive and overbearing" (Kim 2006, p. 547). The emergentist position tries to maintain physicalism and at the same time the irreducibility of mental phenomena arguing that certain highly complex dynamical systems are capable of producing such kind of phenomena given the particular kind of complex causal interactions in such systems. Moreover, they frequently argue that after these emergent properties emerge, they also are capable of having causal effects in the physical world (cf. McLaughlin 1992, p. 51), and these causal effects are different and "autonomous" in respect to their physical "lower level" causal effects. All this makes this kind of position very attractive.

Nonetheless, a great amount of criticism on the idea of strong emergence has been advanced in the literature. First of all, it has been argued that the nature of the dependence relation between this kind of property and its lower-level is still obscure; thus, making the idea of irreducibility appear highly obscure. In the case of Gazzaniga and similar views, no specific dependence relation is advanced and we are left with no explanation at all about how exactly the emergent properties emerge. In the end, it is a matter of accepting it as a brute fact; and not just the rising of the new irreducible properties, but also their autonomous inexplicable causal effects that usually are attributed to such properties. Many authors think that such a position, if it is a form of physicalism at all, does not do any better than forms of dualism, such as substance dualism or property dualism, since all these positions ultimately appeal to a fundamental incomprehensible relation between physical and non-physical phenomena. Moreover, Kim (2006, p. 549) challenges the very possibility of such a position to be considered emergentist at all. According to Kim (2006), classical emergentists accepted mereological supervenience (in which every change in the micro level leads to a change in the macro level). In his view, an emergentist would hardly deny that "if the very same configuration of physiological events were to recur, the same mental phenomenon, [for example], pain, would emerge again" (Kim 2006, p. 550). If such connection were entirely irregular, arbitrary, or just a matter of coincidence, there would be indeed no reason for saying that pain 'emerges from' that neural condition, and not from another. After all, Kim argues, the very meaning of 'neural substrate' implies the presence of some kind of upward determination; if this is taken away, the very idea of emergence appears to be damaged.

In the case a relationship of supervenience is advanced, however, there is as well the difficult problem of mental causal exclusion within the issue of downward causation (causation from a higher-level to a lower-level). According to Kim's popular argument, (1) if a physical event has a cause occurring at time *t*, it has a sufficient physical cause at *t*; and (2) no event has two or more distinct sufficient causes (except in cases of genuine overdetermination, e.g. two or multiple bullets hit the heart of a person at the same time causing death) (2006, pp. 557–558). If these statements are accepted (as generally they are in these debates), together with the thesis of supervenience, it becomes very difficult to understand how strong emergent phenomena can be causally effective in the physical world without being superfluous. To be causally effective and not superfluous, the emergent properties would have to be reduced somehow to the physical realm, but the reducibility of strong emergent phenomena to the physical realm is precisely what the strong emergentist denies (cf. Kim 1998, 1999, 2005, 2006). Accordingly, the strong emergentist has the difficult task of explaining the possibility of a link between a strong emergent phenomenon and its physical basis that entails irreducibility but at the same time maintains the genuine causal effectiveness of that emergent phenomenon (unless one feels satisfied, as few do, with epiphenomenalism, i.e. the thesis that mental phenomena are caused by physical phenomena, but do not cause anything in turn).[21] This is clearly the kind of emergence that the proponents of the mechanistic theory call 'spooky' or 'dualist' and really do not intend to allow in their framework. Therefore, unequivocally this is a card that the mechanists cannot and will not put on the table:

> [...] one who insists that there is no explanation for a non-relational property of the whole in terms of the properties of its component parts-plus-organization advocates a spooky form of emergence. [...] Appeal to strong or spooky emergence [...] justifiably arouses suspicion. (Craver 2007, pp. 216–217)

Another possibility for characterizing emergence is to appeal to its weak form. Weak emergence occurs when a high-level phenomenon arises from the low-level domain in an 'unexpected' way given the complex interactions in the low-level domain. However, it is overwhelmingly pointed out in the literature that this high-level phenomenon *can still be reduced* to the low-level basis. This is clearly contrary to the views of Bechtel and Craver. Contrarily to what they argue, weak emergence suggests that—even though the entities or theories positioned at a higher-level are characterized by some emergent (i.e. novel and unexpected) properties with respect to properties and related processes positioned at a lower-level—the emergent phenomenon can be fully understood and explained in terms of the lower-level components and their relations. One standard example of weak emergence in cognitive science is given by connectionist networks, in which particular kinds of cognitive behavior emerge from single interactions among simple logic units (Chalmers 2006; Stephan 1999). Only trained nets show typical macroscopic properties, that are not present in any of their components (namely in their units and in the links among them), such as 'rule following', 'schemata formation', and 'pattern recognition'. Since the emergent phenomena are completely understood and explained by the lower-level components of the structure of the net, i.e. by its working parts, operations and by the links among its working parts and operations (its organization), the systemic phenomenon of a net are merely weakly emergent. It is thus the very fact that reduction can be applied here that makes weak emergence *weak*. The multilevel mechanists seem to have ignored this. Indeed, their discussion of the notion of 'emergence' is superficial. Accordingly, the weak sense of emergence defended by the mechanistic theory can be considered reductionist. Once again MTHC faces the problem of reduction.

---

[21] There is still a complex ongoing debate in philosophy of mind concerning whether strong emergentists and other nonreductive physicalists can deal with the problem of causal exclusion. At any rate, the strong emergentist position is highly controversial and very difficult to hold.

4.1.5 Organization, Unpredictability, External Factors and Strong Neuroscientific Reduction

The strategy of the most influential advocates of MTHC is, as we saw above, to argue that the 'behavior of a whole mechanism is more than the behavior of its parts'. In their view, since the system is organized in a specific manner, and this organization is a higher-level aspect of the whole mechanism (because it refers to causal processes that are not just related to individual or sets of parts, but rather to the whole system), then there is some independent causal higher-level—consequently, autonomy and plurality of these putative higher-levels.

It is evident that in most cases in nature the operation of just one component, or a set of components smaller than the whole mechanism/system, cannot account for the behavior of the whole. For example, the operations of the motor of a car cannot account alone for the whole motion of the car. Nor taking the motor and the wires together is enough to account for many 'behaviors' of the whole car, i.e. many functions that the entire car can perform. In biological mechanisms such as the human heart, the same occurs: the entire heart performs biological functions within the circulatory system that none of the parts of this heart taken alone can perform. Thus, indeed, a whole often has features that none of its constitutive component parts have and it performs functions that none of its parts alone do.

However, Bickle does not argue that a single component's function, or a set of component functions, always explain the function of a given whole mechanism. This is really a naive view, and Bickle certainly does not take this position, as we saw above. The issue here is that what Bechtel sees as higher-level, for Bickle, can be simply described in lower levels—exactly as weak emergentists claim. Bickle knows that organization is important, but in his view all the information about the organization can be also described in terms of lower-level, i.e. at the level of molecules and cells in the case of brain and cognition, where ultimately lies the most fundamental, detailed and accurate causal description and explanation.

The mechanists assume that the hippocampus and other regions in the human brain form the whole mechanism for memory consolidation, and that this large neural structure performs a particular kind of operation that needs to be understood in order to account for the phenomenon. However, they rely heavily on research concerning LTP, and the LTP processes are best described in terms of chemical synapses, neurotransmitters, molecules and ions. If what is most important for memory consolidation in the hippocampus is LTP, and LTP processes can be described in terms of the activity of molecules in the brain, then we have an explanation of the mechanism of memory consolidation entirely in terms of molecules, synapses and neurotransmitters, just as Bickle argues. The mechanists argue that higher-levels are important because they bring some causal novelty to the picture given the organization (which they argue is neglected by Bickle). Why the component parts in the hippocampal network cannot be simply explained in terms of molecules? There is no answer for this.

The central problem with the mechanistic account is that when its advocates speak about mechanistic *inter-level and multi-level relations*, they emphasize the *compositional relation of part-whole* which is an *asymmetrical* relation, since parts of wholes compose wholes, but wholes do not compose their parts; nevertheless, when they speak about the relation between the *explanandum* and the *explanans*, they emphasize the relation between the whole (and its function under consideration) and the *component parts plus their organization*. But this relation is the *symmetrical* relation of *identity* (or, what is practically the same here, *exhaustive constitution*),[22] since component parts and their organization are exactly what constitutes wholes. It is, therefore, because of the ambiguity present in the framework that it is possible to say, at the same time, that: (1) to consider just a part of a whole is to investigate a lower-level (in this way, to be in a lower-level means *necessarily* to consider just a part of the mechanism); and (2) that what explains the function of a whole is the activity of all its components plus their organization.

Moreover, since good explanations of mechanistic wholes need always to be made considering their parts and their organization, it is wrong *in principle* to seek any kind of explanation that considers just a part of the whole. Thus, Bechtel and others, ultimately, are claiming that Bickle is wrong *in principle* because he is committing a kind of logical or epistemological mistake, since he is trying to explain a whole using just some of its parts, while, by mechanistic explanatory norms, a whole can just be explained by all its parts plus their organization.

Accordingly, the causal novelty putatively present in the hippocampal network, according to the neo-mechanists, does not appear anywhere. Or at least there is no clear evidence presented in support of some novel causality given the organization that is not trivial (i.e. more than just the organizational relation between the causal processes of all the different component parts). Even if episodic memory consolidation is not limited to the hippocampus, what speaks against the possibility of an explanation of the other neural networks involved in term of neurotransmitters, receptors, molecules and ions? This is exactly the kind of explanation that Bickle is trying to provide and this also shows that he does not neglect the organization of a mechanism. Here is what Bickle claims:

> Blocking any step in the cAMP-PKA-CREB process virtually eradicates memory consolidation, while enhancing steps can lead to faster and stronger consolidation. These basic effects have now been demonstrated experimentally for a large number of memory tasks, including hippocampus-dependent 'declarative' or 'explicit' memories. (Bickle 2008, p. 38)[23]

And in another work, he claims this: "[...] the cAMP-PKA-CREB intra-neuron pathway is a molecular mechanism for hippocampus-dependent memory consolidation in mammals" (Bickle 2006, p. 424). As we can note, the author claims the

[22] Cf. Leite (2019).

[23] As Aizawa (2007, p. 68) already noted, Bickle is not so explicitly explaining what exactly he thinks the mechanism of memory consolidation in mammals is. In a passage of his book, he writes that the mechanism includes: "adenylyl cyclase, cAMP, PKA, CREB enhancers and repressors, DNA, RNA polymerases, ubiquitin hydrolase, CCAAT enhancer binding protein, glutamate, dendritic spine cytoskeleton components, AMPA receptors, NMDA receptors, and so on" (Bickle 2003, p. 99, cf. p. 75). It is not clear whether the two accounts are about the same mechanism. At any rate, cAMP-PKA-CREB are certainly indispensable components of it.

cAMP-PKA-CREB mechanism is responsible for generating LTP and thus many kinds of memory consolidation in the hippocampus of mammals. This is the whole mechanism and it is constituted by the processes and activities of its components, namely, those involved in the cAMP-PKA-CREB process. This process happens in synaptic activities all across the hippocampus. Thus, he is not saying that a part, or a sub-part, or a sub-sub-part, etc. of a given mechanism under consideration is explaining the entire mechanism, as Bechtel and Craver argue. Rather, Bickle is saying that the entire mechanism responsible for a very particular function is a molecular-cellular one.

Thus, if the *explanandum* related to the putative mechanisms being compared is not precisely defined, it is hard to make these comparisons. For, since parts and wholes can be considered mechanisms relative to the function (or phenomenon) under investigation, it can easily be the case that one author accepts a given whole as an entire mechanism, while another author accepts this as a mere part of a larger mechanism. It is this ambiguity that makes such a confusion when it comes to the comparison between MTHC and MCTHC.

Ultimately, it appears that the explanatory causal pluralism of MTHC faces some sort of dilemma. For if the novel causal processes assumed by the multilevel mechanists are present just due to the causal processes of all component parts, their activities and organization together, then this is trivial, because this would be the same as explaining all (or most) of the functions of a whole mechanism considering the whole mechanism. But if there is some other kind of causal novelty being defended here by the mechanists, then it is obscure and no empirical evidence or compelling reasons are being given, besides of the ontological and epistemological problems that such account of causation would bring—as the literature in the field of philosophy of mind shows. In this respect, Bickle's framework is much clearer than the mechanistic one.

The same mischaracterization of Bickle's reductionism is repeated in the specialized literature among opponents of reductionism relying on the mechanistic criticism. Chemero and Silberstein (2008, p. 11), for instance, claim that a strictly reductionist view about mechanisms considers that the properties of smaller components of a given mechanism locally determine the properties of the entire mechanism. This shows that, in order to argue against reductionism, not just the proponents of the mechanistic theory mischaracterize the position, but also their followers on this issue, creating a straw man and transforming it in a very easy target, as Fazekas and Kertész (2011) correctly point out.

Interestingly, the triviality in this allegedly form of reductionism was already pointed out in the classic literature many years ago. The philosopher Mario Bunge published a paper, in 1977, titled *Emergence and the Mind*. In this paper he distinguishes a 'resultant property' from an 'emergent property'; the resultant property of a given system is a property of some components of the system, and an emergent property is a property just of the entire system, in which no component part has the particular property (Bunge 1977, p. 502). In a discussion of Bunge's proposal, Stephan mentions "a trivial type of reductionism which would explain properties of systems by ascribing them to the parts of it" (1992, p. 32). And Stephan goes on to say that he knows of no one defending this kind of reductionism in the literature. So far, this has not changed. There is no recognized author in the specialized literature who advances a kind of reductionism claiming that certain parts of a given system are capable of explaining the macro properties or functions of the entire system.

Even in cases concerning complex systems, where there is high unpredictability, there is reduction. But, before addressing this point, let us start with some clarifications.[24] Unpredictable systems, or unpredictable properties of systems, can indeed emerge given the indeterminism of the physical world. The theory of chaos states that "there exist mathematical functions, whose own 'behavior' cannot be predicted"; this mathematical unpredictability "has to do with aperiodic behavior of these functions, by which marginally different initial values of some variable can lead to radically distinct trajectories of the functions" (Stephan 1999, p. 54). Thus, the theory of chaos suggests the existence of unpredictable systems. It is still a debated issue, nevertheless, if chaotic systems cannot be predicted in principle. One could argue that if a rational creature could possibly know all the previous states of the world and all the processes and laws governing them, this creature could predict the appearance and features of all the future physical structures. At any rate, no cognitive scientist or neuroscientist (indeed no scientist) has this wonderful capacity. Thus, "where chaos exists, structures exist that are unpredictable in principle" (Stephan 1999, p. 54).

Craver's and Bechtel's concern is with cognitive properties that are unpredictable given a complex neural micro-structure and its organization. However, first of all, it is not clear if indeterministic chaos plays a major role, or even any role at all in neuroscience. At any rate, nevertheless, contrarily to their point of view, reduction can stand even despite of such unpredictability. There is indeed no conflict between unpredictability in complex systems and Bickle's reductionism. It is overwhelmingly accepted in the literature on weak emergence that this kind of unpredictability is hardly a problem for reduction. There is no good reason why reduction would necessarily require a high degree of predictability. For it is irrelevant if at a certain point in time a phenomenon or a set of phenomena occur that cannot be completely predicted with great accuracy on the basis of the initial state of the system due to the inaccuracy of the measures at the start. What really matters for Bickle's strong neuroscientific reduction is whether it is possible to describe these phenomena in causal lower-level mechanistic physical language—and then to identify as many mechanistic variables as possible, put all the information together and ultimately produce a scientific explanation of the whole complex system's behavior at hand. This does not mean that one should give up on predictability once and for all, of course. It just means that the more dynamical and complex a system is, the less its outcomes will be predictable given the incommensurable number of variables and complex causal relations between them.

---

[24]Bechtel and Craver are not the first authors to mention predictability as a characteristic of emergent properties and other contemporary authors follow the same idea (e.g. Gazzaniga 2012). However, none of these authors mention the roots of this thesis. For example, the work of the philosopher Karl Popper, in his book of 1977, *The Self and Its Brain*, published together with John Eccles. Popper is considered to be "one of the strongest proponents" of non-predictability of emergent properties (Stephan 1992, p. 32). But besides Popper, Lloyd Morgan and C. D. Broad had already used unpredictability as a criterion for emergence.

In fact, all examples of complex systems in nature can be easily described in reductive terms, in spite of the unpredictability involved. Weather and ecosystems are dynamical complex systems that have features which cannot be predicted with complete accuracy especially across longer time spans. Nevertheless, one could easily describe all the component parts of the system in lower-level mechanistic physical terms through the functions and physical causal roles of the system and its components, subcomponents and so on and so forth. Such examples are completely reducible in this sense even if some of them are under certain conditions unpredictable. Accordingly, unpredictability does not entail irreducibility as long as the appropriate reductive theory is considered. In this reductive picture the relevant processes required for the final explanation can be acquired from the lower level of the component operations and their interactions; then this information can be put together in order to explain all the causal interactions at the higher-level, since they would be equivalent.

Another related point worth to mention concerns the fact that unpredictability here is strictly related to the idea of non-deducibility. However, proponents of the mechanistic theory often argue against the notion of 'deduction of laws', related with Nagel's intertheoretical theory of reduction and with the DNTSE. In this sense, they are also arguing against predictability so considered. Thus, it does not make much sense to argue against (unqualified) unpredictability as a problem for reduction if one already took Nagel's classic theory of reduction out of the table. The sciences of life are not committed to a high degree of predictability based on universal laws as it occurs in physics. One does not say, based on this, that the phenomena in biology which cannot be predicted with a high degree of accuracy, such as the flying behavior of a group of birds, or the development of a given ecosystem, are irreducible in some interesting explanatory sense. The neo-mechanists, therefore, when arguing against reduction using unpredictability mixture two very different reductionist frameworks, namely, Nagel's one and Bickle's one. While for Nagel prediction is indispensable, for Bickle it is important, but does not play a major role. Indeed, systems can be well understood and explained, but at the same time they can be unpredictable: as weather or hurricanes. Similarly, systems can be predicted and be poorly understood and explained: one can predict what is going to happen by systematically observing a regular connection in nature without having any idea about why it happens in that way. Thus, explanation and prediction are also two different things and one does well by not merging them together in these debates. Bickle's framework does not require high level of predictability in neural systems and he does not need to. There can be Bickle reduction even if systems are highly unpredictable.

Finally, the last major line of answer that the neo-mechanists provide, as we saw before, is to appeal to the external factors that arguably play a causal role at the higher-level of the whole mechanism in order to determine its behavior. However, there are also serious problems with this answer. Since the interactions at the higher-level are also among mechanisms, i.e. the whole target mechanism and its environment, which is composed of other mechanisms, there is nothing that prevents all these whole mechanisms that interact to be further decomposed in sub-components and sub-functions. There is again a mischaracterization of reduction here, as if the reduction to lower levels could not be equally applied to whole mechanisms that interact with a target whole mechanism at this high level of wholes (note that the hierarchy of levels is always relative, since it depends on the phenomenon being target). This point was correctly noted by Fazekas and Kertész (2011, p. 378). The authors point out that everything that is in the high-level context of a whole target mechanism can be further decomposed and explained in lower levels. Since they are all mechanisms, the reduction should be applied to all of them, without exception. Indeed, Bechtel and Craver give no reason to think about exceptions here.

To sum up, no author claims that one single component of a given whole mechanism will be doing the causal work alone and that this is sufficient to explain the phenomenon performed by the entire system. Everyone agrees that eventually all the components and their operations must be put together according to their organization to account for the behavior of the whole mechanism, which is always the target *explanandum*. Besides, not just the total sum of internal operations of the components of a mechanism are important, but also the relationship the given whole mechanism has with external factors (environment; external stimuli). Bechtel wants to give the impression that reduction means "appealing to lower levels to explanation" while anti-reductionists appeal to higher-levels (2008, p. 143). In this way, he can offer a straightforward solution, namely, lower levels and higher-levels in the same picture—reduction and autonomy at the same time. In other words: "By accommodating both a reductionist and an emergentist perspective, mechanistic explanation provides a unifying framework that integrates a variety of explanatory projects in psychology" (Bechtel and Wright 2009, p. 127).

However, to defend reduction does not mean simply to say that lower levels are important; and anti-reductionists do not claim just that higher-levels are important. Equating decomposition with reduction does not solve any genuine problem.[25] In fact, decomposition has nothing to do with it. Many anti-reductionists can agree that decomposition is important, without agreeing that it is reductive. As we saw before, theories of reduction are established because they help the understanding of scientific theoretical integration and progress. But in which sense Bechtel's remarks on decompositional reduction help us to understand scientific theoretical development in any sense that is not trivial? Besides, no serious account of reduction claims that there are specific components of mechanisms that can fully account for the behavior of the whole. There is, thus, no theory of reduction that makes sense grounding any part of the mechanistic explanation. The idea of reduction presented by Bechtel is, first of all, not clear and well-articulated; secondly, it is completely dispensable for explaining any part of the scientific activity. Reduction here, hence, is retained just as a word without any theoretical relevance.

The important issue here concerns whether the relevant causal interactions present in the final explanation will have to be described according to multiple levels, or can be simply described in the molecular and cellular lowest one. The real issue is where ultimately the most important and accurate causal explanation lies. In the neo-mechanists' view, the organization of all the components operations inside the

[25] Bechtel himself recognizes that his account of mechanistic reduction and the autonomy of sciences that come with it are "weak" (2007, p. 174).

mechanism plus the external factors that affect the mechanism are all important for the ultimate causal explanation. In this way, they seek to argue against reduction. However, as we saw, although the mechanistic theory is clearly committed to causal pluralism instead of any genuine account of reduction, it is not able to provide a clear defense of it against Bickle's genuine framework of strong neuroscientific reduction.

The analysis show, therefore, that the pluralist aspirations present in the mechanistic theory are indeed effectively challenged by the strong reductionist position.[26] The proponents of the mechanistic framework suggest that their view is a kind of weak emergentist position: a middle term between ruthless reductionism and 'spooky' emergentism, vitalism or any other kind of dualism. However, the interpretations of ruthless reductionism and weak emergentism made by the advocates of MTHC present shortcomings. Ruthless reductionism is not so extreme and implausible as they try to argue and weak emergentism can be seeing as completely compatible with reduction. Indeed, the mechanistic theory's proposal is an attempt to develop some kind of weak emergentism and to this extent it collapses into a form of reduction. Thus, the situation here is not the one in which we have two extreme and implausible positions and a middle one more reasonable and plausible trying to propose an alternative. Rather, the influential neo-mechanists attempt to find space where there is none, misinterpreting the alternatives. This constitutes a major obstacle for their pluralist explanatory ambitions.

4.2 Final Remarks

I analyzed, in this chapter, central aspects of a major alternative research approach to MTHC, together with the challenges it advances for the neo-mechanist's framework. The central purpose of the analysis was to understand to what extent MTHC is able to deal with the challenges pointed out by its competitor and provide better views on the particular issues discussed here. This major alternative approach, MCTHC, is most strongly represented by the work of John Bickle, and the major challenge it poses to MTHC is that it advances a strong reductionist account, instead of a pluralist account, to the scientific integration of the research fields interested in human cognition. It is thus a challenge to one of the most central theoretical aspects of the mechanistic framework.

The analysis shows that the neo-mechanists' plea for plurality fails when it meets Bickle's reductionism; and although proponents of MTHC intend to avoid a strong commitment to reduction, the mechanistic framework indeed collapses into a form of strong reductionism. The most central arguments the advocates of the mechanistic theory use against reduction are based on mischaracterizations. They argue that the reductionist position is committed to the view that 'a part of a whole must explain

[26] In the specialized literature, many authors have also correctly considered contemporary mechanistic explanations as reductionist, e.g. Fazekas and Kertész (2011), Soom (2012), Theurer (2013) and Rosenberg (2015).

the behavior of the entire whole'. That is not correct, however. Serious reductionists are well aware that individual parts do not explain the behavior of the whole (at least in the life sciences); all they say is that all the aspects of such explanation can be achieved at a lower level. This includes the organization of the whole, includes systems with high degree of complexity, and includes the external factors affecting the system, as the specialized literature on weak emergence also clearly recognizes. Consequently, no clear form of causal mechanistic plurality and explanatory plurality can be achieved in such lines. In this regard, it is important to note that my intention in this chapter is not to argue for, or defend, reduction against MTHC. My aim is solely to point out to an inconsistency in the theory, namely, the fact that it is presented as a kind of pluralist theory, when ultimately it is no more than a kind of strong reductionist one.

This analysis shows, therefore, that the problems that neuroscientific reduction poses are not yet overcome. This is indeed a very negative outcome for MTHC. The indecision towards reductionism or pluralism makes it look self-contradictory and inconsistent. Given the most frequent attempts at arguing against reduction by neo-mechanists, however, one would expect that MTHC could indeed be regarded as a pluralist theory. But after a rigorous and systematic analysis, one finds out that the arguments are indeed weak and that the theory simply collapses into a reductionist view. As a result, one possibility for the theory is to make indeed the clear defense of what it really does: namely, reduction. In this sense, the theory loses in terms of ambitions, but wins in terms of clarity and consistency.

References

Aizawa, K.: The biochemistry of memory consolidation: A model system for the philosophy of mind. Synthese **155**, 65–98 (2007)

Bechtel, W.: Levels of description and explanation in cognitive science. Mind. Mach. **4**, 1–25 (1994)

Bechtel, W.: Reducing psychology while maintaining its autonomy via mechanistic explanations. In Schouten, M., Looren de Jong, H. (eds.) The Matter of the Mind: Philosophical Essays on Psychology, Neuroscience, and Reduction (pp. 172–198). Malden, MA: Blackwell Publishing (2007)

Bechtel, W.: Mental Mechanisms: Philosophical Perspectives on Cognitive Neurosciences. Routledge, New York (2008)

Bechtel, W.: Molecules, systems, and behavior: Another view of memory consolidation. In: Bickle, J. (ed.) The Oxford Handbook of Philosophy and Neuroscience, pp. 13–40. Oxford University Press, New York (2009)

Bechtel, W., Abrahamsen, A.: Explanation: A mechanist alternative. Stud. Hist. Philos. Biol. Biomed. Sci. **36**, 421–441 (2005)

Bechtel, W., Richardson, R.: Emergent phenomena and complex systems. In: Beckermann, A., Flohr, H., Kim, J. (eds.) Emergence or Reduction? Essays on the Prospects of Nonreductive Physicalism, pp. 257–288. Walter de Gruyter, Berlin (1992)

Bechtel, W., Wright, C.D.: What is psychological explanation? In: Symons, John, Calvo, Paco (eds.) The Routledge Companion to Philosophy of Psychology, pp. 113–130. Routledge and Taylor & Francis Group, New York (2009)

Beckermann, A., Flohr, H., Kim, J. (eds.): Emergence or Reduction? Essays on the Prospects of Nonreductive Physicalism. Walter de Gruyter, Berlin (1992)
Bedau, M., Humphreys, P. (eds.): Emergence: Contemporary Readings in Philosophy and Science. MIT Press, Cambridge (2008)
Bickle, J.: Philosophy and Neuroscience: A Ruthlessly Reductive Account. Kluwer Academic Press, Norwell, MA (2003)
Bickle, J.: Reducing mind to molecular pathways: Explicating the reductionism implicit in current cellular and molecular neuroscience. Synthese **151**, 411–434 (2006)
Bickle, J.: Real reduction in real neuroscience: Metascience, not philosophy of science (and certainly not metaphysics!). In: Hohwy, J., Kallestrup, J. (eds.) Being Reduced: New Essays on Reduction, Explanation, and Causation, pp. 34–51. Oxford University Press, Oxford (2008)
Bickle, J.: A brief history of neuroscience's actual influences on mind–brain reductionism. In: Gozzano, S., Hill, C.S. (eds.) New Perspectives on Type Identity: the Mental and the Physical, pp. 88–110. Cambridge University Press, Cambridge (2012)
Bickle, J.: Marr and reductionism. Top. Cogn. Sci. **7**, 299–311 (2015)
Bickle, J.: Revolutions in neuroscience: Tool development. Front. Syst. Neurosci. (2016). https://doi.org/10.3389/fnsys.2016.00024
Bunge, M.: Emergence and the mind. Neuroscience **2**, 501–509 (1977)
Carnap, R.: The Unity of Science. Kegan Paul, Trench, Trubner, and Co, London (1934)
Chalmers, D.: The Conscious Mind. Oxford University Press, Oxford (1996)
Chalmers, D.: Strong and weak emergence. In: Clayton, P., Davies, P. (eds.) The Re-emergence of Emergence, pp. 244–256. Oxford University Press, Oxford (2006)
Chalmers, D., Jackson, F.: Conceptual analysis and reductive explanation. Philos. Rev. **110**, 315–360 (2001)
Chemero, A., Silberstein, M.: After the philosophy of mind: Replacing scholasticism with science. Philos. Sci. **75**(1), 1–27 (2008)
Churchland, P.S.: Neurophilosophy. MIT Press, Cambridge, MA (1986)
Churchland, P.S.: Can neurobiology teach us anything about consciousness? Proc. Addresses Am. Philos. Assoc. **67**, 23–40 (1994)
Churchland, P.S., Sejnowski, T.: The Computational Brain. MIT Press, Cambridge, MA (1992)
Corradini, A., O'Connor, T. (eds.): Emergence in Science and Philosophy. Routledge, New York (2010)
Crane, T.: The significance of emergence. In: Gillett, C., Loewer, B. (eds.) Physicalism and Its Discontents, pp. 207–224. Cambridge University Press, Cambridge (2001)
Craver, C.F.: Beyond reduction: Mechanisms, multifield integration and the unity of neuroscience. Stud. Hist. Philos. Biol. Biomed. Sci. **36**, 373–395 (2005)
Craver, C.F.: Explaining the Brain: Mechanisms and the Mosaic Unity of Neuroscience. Oxford University Press, Oxford (2007)
Crick, F.: The Astonishing Hypothesis: The Scientific Search for the Soul. Touchstone and Simon & Schuster, New York (1995)
Edelman, G.: The Remembered Present. Basic Books, New York (1989)
Edelman, G.: Wider Than the Sky: The Phenomenal Gift of Consciousness. Yale University Press, New Haven (2004)
Fazekas, P., Kertész, G.: Causation at different levels: tracking the commitments of mechanistic explanations. Biol. Philos. **26**, 365–383 (2011)
Feigl, H.: The "mental" and the "physical". In: Feigl, H., Scriven, M., Maxwell, G. (eds.), Concepts, Theories and the Mind-Body Problem (Minnesota Studies in the Philosophy of Science, Volume 2), University of Minnesota Press, Minneapolis (1958)
Gazzaniga, M.S.: The Bisected Brain. Appleton-Century-Crofts, New York (1970)
Gazzaniga, M.S.: Social Brain: Discovering the Networks of the Mind. Basic Books, New York (1985)
Gazzaniga, M.S.: Mind Matters: How Mind and Brain Interact to Create our Conscious Lives. Houghton Mifflin, Boston (1988)

Gazzaniga, M.S.: Nature's Mind: The Biological Roots of Thinking, Emotions, Sexuality, Language and Intelligence. Basic Books, New York (1992)
Gazzaniga, M.S.: The Mind's Past. University of California Press, Berkeley (2000)
Gazzaniga, M.S.: The Ethical Brain. Dana Press, New York (2005)
Gazzaniga, M.S.: Tales from Both Sides of the Brain: A Life in Neuroscience. Ecco (Harper-Collins), New York (2015)
Gazzaniga, M.S.: Who's in Charge? Free Will and the Science of the Brain, 2nd edn. Ecco (Harper-Collins), New York (2012)
Gazzaniga, M.S.: Human: The Science Behind What Makes Us Unique. Ecco (Harper-Collins), New York (2008)
Gazzaniga, M.S., LeDoux, J.E.: The Integrated Mind. Plenum Press, New York (1978)
Hooker, C.: Towards a general theory of reduction. Part I: Historical and scientific setting. Part II: Identity in reduction. Part III: Cross-categorial reduction. *Dialogue* **20**, 38–59, 201–236, 496–529 (1981)
Kemmeny, J.G., Oppenheim, P.: On reduction. Philos. Stud. **7**(1/2), 6–19 (1956)
Kim, J.: Mind in a Physical World: An Essay on the Mind-Body Problem and Mental Causation. MIT Press, Cambridge, MA (1998)
Kim, J.: Making sense of emergence. Philos. Stud. **95**, 3–36 (1999)
Kim, J.: Physicalism, or Something Near Enough. Princeton University Press, Princeton (2005)
Kim, J.: Emergence: Core ideas and issues. Synthese **151**(3), 347–354 (2006)
Leite, D.A.: Neo-mechanistic explanatory integration for cognitive science: The problem of reduction remains. Sofia **8**(1), 124–145 (2019)
Levine, J.: Materialism and qualia: The explanatory gap. Pac. Philos. Q. **64**, 354–361 (1983)
Levine, J.: On Leaving out what it's like. In: Humphreys, G., Davies, M. (eds.) Consciousness: Psychological and Philosophical Essays, pp. 121–136. Blackwell, Oxford, UK (1993)
Macdonald, C., Macdonald, G. (eds.): Emergence in Mind. Oxford University Press, Oxford (2010)
McLaughlin, B.: The rise and fall of British emergentism. In: Beckermann, A., Flohr, H., Kim, J. (eds.) Emergence or Reduction? Essays on the Prospects of Nonreductive Physicalism, pp. 49–93. Walter de Gruyter, Berlin (1992)
Nagel, E.: The Structure of Science: Problems in the Logic of Explanation. Harcourt, Brace & World, Inc, New York (1961)
O'Connor, T.: Emergent properties. Am. Philos. Q. **31**, 91–104 (1994)
O'Connor, T., Wong, H.Y.: Emergent Properties. The Stanford Encyclopedia of Philosophy (Summer 2015 Edition). In Zalta, E.N. (ed.) (2015). http://plato.stanford.edu/archives/sum2015/entries/properties-emergent/
Place, U.T.: Is consciousness a brain process? Br. J. Psychol. **47**, 44–50 (1956)
Ramachandran, V.S.: The Tell-Tale Brain: A Neuroscientist's Quest for What Makes Us Human. W. W. Norton, New York (2011)
Reuter-Lorenz, P.A, Baynes, K., Mangun, G.R., Phelps, E.A. (eds.).: *The Cognitive Neuroscience of Mind: A Tribute to Michael S. Gazzaniga*, MIT Press, Cambridge (2010)
Richardson, R.C., Stephan, S.: Reductionism (antireductionism, reductive explanation). In: Binder, M., Hirokawa, N., Windhorst, U. (eds.) Encyclopedia of Neuroscience, pp. 3395–3398. Springer, Heidelberg (2009)
Rosenberg, A.: Making mechanism interesting. Synthese (2015). https://doi.org/10.1007/s11229-015-0713-5
Ryle, G.: The Concept of Mind. Hutchinson, London (1949)
Sarkar, S.: Models of reduction and categories of reductionism. Synthese **91**(3), 167–194 (1992)
Schaffner, K.: Approaches to reduction. Philos. Sci. **34**, 137–147 (1967)
Silberstein, M.: Reduction, emergence, and explanation. In: Silberstein, M., Machamer, P. (eds.) The Blackwell Guide to the Philosophy of Science, pp. 80–107. Blackwell, Oxford, UK (2002)
Silva, A., Bickle, J.: The science of research and the search for molecular mechanisms of cognitive functions. In: Bickle, J. (ed.) The Oxford Handbook of Philosophy and Neuroscience, pp. 91–126. Oxford University Press, New York (2009)

Silva, A.J., Landreth, A., Bickle, J.: Engineering the Next Revolution in Neuroscience. Oxford University Press, New York, NY (2014)
Smart, J.: Sensations and brain processes. Philos. Rev. **68**(2), 141–156 (1959)
Soom, P.: Mechanisms, determination and the metaphysics of neuroscience. Stud. Hist. Philos. Sci. **C43**, 655–664 (2012)
Soon, C.S., Brass, M., Heinze, H.J., Hayes, J.D.: Unconscious determinants of free decisions in the human brain. Nat. Neurosci. **10**, 257–261 (2008)
Stephan, A.: Emergence—A systematic view on its historical facets. In: Beckermann, A., Flohr, H., Kim, J. (eds.) Emergence or Reduction? Essays on the Prospects of Nonreductive Physicalism, pp. 25–48. Walter de Gruyter, Berlin (1992)
Stephan, A.: Armchair arguments against emergentism. Erkenntnis **46**, 305–314 (1997)
Stephan, A.: Varieties of emergentism. Evolution and Cognition **5**(1), 49–59 (1999)
Stephan, A.: Emergentism, irreducibility, and downward causation. Grazer Philosophische Studien **65**, 77–93 (2002)
Stephan, A.: The dual role of 'emergence' in the philosophy of mind and in cognitive science. Synthese **151**, 485–498 (2006)
Stephan, A.: Emergence, theories of. In: Runehov, A., Oviedo, L. (eds.) Encyclopedia of Sciences and Religions, pp. 714–721. Springer, Dordrecht (2013)
Theurer, K.L.: Compositional explanatory relations and mechanistic reduction. Mind. Mach. **23**, 287–307 (2013)
Van Gulick, R.: Reduction, emergence and other recent options on the mind/body problem: A philosophic overview. J. Conscious. Stud. **8**(9–10), 1–34 (2001)
van Riel, R., Van Gulick, R.: Scientific Reduction. The Stanford Encyclopedia of Philosophy (Fall 2015 Edition). In Zalta, E.N. (ed.), forthcoming (2014) http://plato.stanford.edu/archives/fall2015/entries/scientific-reduction/
Walter, S., Eronen, M.: Reductionism, multiple realizability, and levels of reality. In: French, S., Saatsi, J. (eds.) Continuum Companion to the Philosophy of Science, pp. 138–156. Continuum, London (2011)

Chapter 5
Dynamical Systems Theory of Human Cognition and Theory of Situated Human Cognition

5.1 DSTHC Against MTHC

5.1.1 The Challenge to Decomposition and Localization

The use of complex dynamical systems in debates concerning cognition goes far beyond the simple mention of examples of such systems to make particular points about the issues of reduction or emergence. There is indeed an articulated framework for understanding cognition in general and human cognition particularly grounded in the complex systems theory. I call this theory in its application to human cognition *Dynamical Systems Theory of Human Cognition* (DSTHC).

In the 1990s, some authors argued for scepticism concerning the classical computational approach in cognitive science. In their view, cognition is to be understood not in terms of computations over representations, but rather as complex dynamical systems, which can be described and completely explained using the concepts and methods of dynamical systems theory (Chemero 2009; Van Gelder 1995, 1998). In present days, dynamical explanations are frequently present in cognitive science and many domains of neuroscience (Kaplan and Bechtel 2011, p. 439). Authors working with dynamical explanations in cognitive science are considered to be a substantial and growing minority (Stepp et al. 2011, p. 435).

Complex dynamical systems are systems with a large number of variables that interact causally in very complex ways, e.g. in a non-linear form and with positive and negative feedback loops (cf. Chap. 3, Sect. 3.1.2). Such systems also change through time in a systematic way and this change can be then described and analysed in a mathematically precise way by means of the mathematical theory of dynamical systems (a part of theoretical mathematics), which originated in the work of Henri Poincaré (1854–1912) and has been further developed from the twentieth century until present days.

The characteristics of a system that change over time serve as the variables, parameters, that determine the total state of a system, and the set of all possible states of the system is its so-called 'state' or 'phase space'. Since the behavior of a

D. A. Leite, *The Twenty-First Century Mechanistic Theory of Human Cognition*, Cognitive Systems Monographs 41, https://doi.org/10.1007/978-3-030-63680-7_5

system is nothing but a series of state changes, it can be described by means of a set of equations, usually differential equations, as a trajectory (path) through its state space. Thus, in principle, it is possible to describe and predict the behavior of the complex system.

Cognitive systems seem to undergo continuous changes that cannot be reduced to algorithmically specifiable sequences of discrete computational operations. In the brain, for instance, there is arguably no strict succession of sequential operations either, but a continuously changing and intricate interaction of bottom-up- and top-down-processes. The idea of more or less rigid linear 'sense-think-act' processes is also problematic, given that sensing and acting seem to be interwoven much more intimately. The interactions between different cognitive subsystems through time appear in this way to be more important than internal computational operations. Therefore, the human cognitive system and its subsystems can be best characterized as complex dynamical systems, in the view of dynamicists. Moreover, the human cognitive system seems to manifest the self-organization also present in dynamical systems. There is no control unit, but the control of the system comes from its very organization (for example, the stepping behavior of infants that results from the self-organizing interaction of different aspects of the child's bodily activity, from its memories of past movement patterns, its posture, the variety of the stimulus etc.).

The major challenge that this framework poses to the neo-mechanistic theory is that, in the view of the dynamicists, in neuro-cognitive complex systems the component parts cannot be understood in isolation. Therefore, there can be no decomposition and localization. Every time a variable is taken out of the system in order to be understood it loses automatically the relevant information concerning the role it plays in the system. In this sense, the relevant large-scale neural dynamics together with the environmental variables that affect them provide a uniform explanation that account for the whole (cognition, brain, body, and environment). Perception and cognition here, for instance, are not understood as completely different systems that occasionally interact with each other in a simple linear way. Contrarily, they are understood as inseparable components in an integrated system that interacts causally in a non-linear complex way.

Chemero and Silberstein (2008, p. 16), indeed, claim that "nonlinearity is bad news for mechanistic explanation", because in the case of nonlinear systems, where the behaviors of the components of the systems are highly correlated, decomposition will fail; in cases such as these, the system cannot be treated as a collection of isolated parts, and these parts are best described in terms of macro structures of the system. So, the more nonlinear a system is, the more localizability and decomposability will fail (Chemero and Silberstein 2008, p. 18).

In the view of Chemero and Silberstein (2008), thus, localization and decomposition are features that most centrally characterize mechanistic explanations. If these explanatory procedures cannot be performed, the entire proposal of the mechanists will fall. With nonlinearity, so they argue, these two tenets cannot be achieved in the explanation of such systems. For those cognitive processes that are best described in nonlinear ways, thus, the mechanistic theory cannot be applied. Accordingly, if all cognitive processes are complex dynamical processes, the mechanistic explanation will fail for each and every one. It is, though, still an open question if the brain really works completely, partially, or at all in a nonlinear way. However, with the progress of scientific understanding about specific neural systems in the brain, scientists are becoming more and more sceptical about the possibility of a simple linear causal explanation of such neural complexity. There is at least a great amount of evidence of systems in the brain working in a very integrated way, not as isolated systems.

The dynamicists argue further that since this decomposition is impossible in certain cases, this means that reduction is out of the picture (cf. Chemero and Silberstein 2008, p. 8). In this case, mechanists fail, on the one hand, because they lose precise decomposition and localization, but get what they want, on the other hand, since this would mean explanatory antireductionism, and thus weak autonomy and plurality of higher-levels. As the dynamicists argue "the relative failure of localizability and decomposability in self-organizing systems implies that the deepest explanation for such systems cannot be in terms of the Lego-philosophy of atomism or mechanism" (Chemero and Silberstein 2008, p. 19). This means for them that the explanations of complex dynamical systems will necessarily conflict with reductionism. This explanatory reductionism, as dynamicists view it, consists in the claim that there are component parts which are independent of context and that all the higher-level features of the system will be determined by the fundamental constituents—the higher-level explanation will follow given information acquired through the investigation of the lower-level (Chemero and Silberstein 2008, p. 19).

However, for the dynamicists, complex dynamical systems, contrarily, are not structured aggregates of physical parts; rather, complex dynamical patterns "emerge from the collective behavior of coupled elements in a particular context and in turn the behavior of each individual element is constrained by the collective behavior of the whole in that context" (Chemero and Silberstein 2008, p. 19). In a typical complex dynamical system, the 'emergent features' of the system arise from the behavior of the coupled elements, yet also constrain their behavior.[1] In their view, in a highly complex dynamical system, the emergent features or behavior of the whole system will cause something in the parts of this system.

As a result, the kind of causal pluralism and ontological holism found in complex dynamical systems presents a problem to this kind of explanatory reductionism (Chemero and Silberstein 2008, p. 20). Thus, mechanistic explanations achieved through decomposition and localization is rejected as incompatible with dynamical explanations and incapable to account for the kind of complex phenomena that are present in cognitive science, neuroscience and other sciences.

[1] Note that the dynamicists also use the concept of 'emergence'.

5.1.2 The Mechanists' Reply: Embracing Complexity

The clearest and strongest reply given by the advocates of the neo-mechanistic theory is the claim that inquiry in complex dynamical system does not undermine mechanistic decomposition and localization, but it is rather compatible with this kind of explanation. Furthermore, it is also argued that decomposition and localization provide indeed the very foundation for dynamical analysis (Bechtel 1998, pp. 295, 309, 2001, pp. 483, 484; Kaplan and Bechtel 2011, p. 438).

The crucial point here is that the mechanists emphasize the organization of the system and do not make any specific requirements on the kind of organization that a given system needs to have. In Bechtel's view, the discovery of complex interactions between components of a given system is a common outcome of mechanistic explanations (2001, p. 484). He emphasizes that the only requirement is to get a first approximate model on how the component parts work by examining them individually, and then to consider their interactions (Bechtel 2001, p. 485). After all, components cannot be considered components in a particular system if they do not participate in appropriate interactions. Some of these interactions are simple and linear: when A affects B and B, in turn, affects C. Other interactions are complex: when A affects B, but is affected back in turn and this changes the way it affects B in the next time, and this interaction also affects the way A and B affect C, D, E, F and so on. In this last case, the components of the system become affected by the interaction within the system and their operations cannot be understood anymore in complete isolation from the system. In that specific environment, the particular components operate constrained by the general conditions; outside that environment, they might operate differently.

However, this does not mean, according to the neo-mechanists, that decomposition and localization are completely out of the picture. First of all, although the brain appears to work in a very complex, dynamical and integrated way, there are different areas in the brain, cortical and subcortical, different types of neurons, and different types of neural connections that can be determined. In present days, decomposing the brain in specific areas is a common practice and very useful. As Bechtel emphasizes, that is a common strategy in cognitive science as well as in much of contemporary neuroscience (1998, p. 308). In these fields, researchers try to decompose the functions performed by the neural systems into subfunctions and then attempt to provide evidence that these functions are performed by particular neural components. For instance, the visual system of macaque monkeys is considered by some authors to have 32 different areas (Felleman and Van Essen 1991) and these areas are linked to different visual functions. The visual system of animals is therefore very complex; however, there are distinct areas carrying out different operations and these areas through their operations interact with each other in a very structured manner (Bechtel 2001, p. 487). At any rate, given the vast number of components, operations and different interconnections in the visual system, dynamical analysis can also be helpful. There is no incompatibility here, no conflict between explanations, but rather a complementation.

It is similar with memory (Bechtel 2001, p. 487ff). This complex phenomenon is related to many large areas in the brain such as the hippocampus, temporal and frontal cortex. There are many subdivisions concerning memory processes (cf. Chap. 3, Sect. 3.2.2). It has been intensively debated what the relations between these kinds of memory are and how exactly they relate to the brain areas. It is very likely that the final explanation will contain many highly complex causal interactions between all the components of the memory system and they are probably many. In this case, dynamical explanations can play an important role. However, at the same time, it is important to provide different analysis to some of the most central different components. Therefore, dynamical and mechanistic explanations can be viewed as complementary on research concerning memory as well.

A final example of how mechanistic and dynamic explanations can be applied together is found in the "Hodgkin-Huxley theory of the action potential", which is "both mathematical-dynamical, comprising a set of coupled differential equations to describe the dynamics of the membrane potential" and at the same time "a mechanistic explanation describing how the components (ion channels) and the activities involving these channels are organized and orchestrated to generate action potentials" (Kaplan and Bechtel 2011, p. 439).

Moreover, in the view of the neo-mechanists, proponents of dynamicism often equivocally equate good explanations with the predictive power that their dynamical tools allow them to provide in connection with the complex behavior of a given system through time. Dynamicists believe that the more predictive power a given explanation has, the better it is. Besides, they believe that decomposition and localization are irrelevant for the prediction of dynamical systems' behaviors; what is relevant for them is the analysis in terms of dynamical systems equations concerning the whole system. In the mechanist view, this is wrong because explanatory adequacy is not mere predictive adequacy. Neo-mechanists argue that the occurrence of a storm can be predicted by knowing a law-like regularity related to the falling mercury in the barometer, and thus the fall of temperature; but the falling mercury cannot be used to explain the occurrence of the storm (Kaplan and Bechtel 2011, p. 440). In the case of a cognitive phenomenon, a dynamical explanation could provide precise and accurate predictions, but the relevant variables could be just mere correlates of a common cause. Thus, dynamical explanations do not really explain in this case, whereas mechanistic explanations explain, because they point out to what the causal interactions in a given mechanism that produce the cognitive phenomenon being investigated are. Moreover, it is possible to improve the accuracy and precision of a given explanation providing more details concerning the component parts and interactions in a given system, but this does not necessarily lead to an increase in the predictive power of the model. For these reasons, in the view of the neo-mechanists, good explanations must provide an account of the most important components and interactions within a system. And this is provided by the mechanistic framework. Predictions are important, but not sufficient for good explanations.

With these arguments, the advocates of the mechanistic theory do not intend to deny that dynamical explanations can be very useful in cognitive science and neuroscience. Instead, they intend to emphasize that without some inquiry on the component parts and operations of the system, dynamical explanations are incomplete or even empty (Kaplan and Bechtel 2011, p. 443). In this respect, thus, dynamical explanations do not offer a new revolutionary paradigm to understand cognitive processes; indeed, just a complement to the neo-mechanistic framework.

Therefore, the solution for the neo-mechanistic framework concerning the issues of decomposition and localization is rather simple. It just requires from the mechanistic theory to allow that at a certain level of complexity, when the variables are too many and the causal interactions too complex and not linear, decomposition is not possible anymore, and could be even unnecessary for the explanation, as long as some reasonable accuracy concerning the causal processes and the prediction is achieved in the end. Thus, at higher-levels of complexity, mechanistic explanation becomes also dynamical and there is no contradiction in this, since the theory is not committed to decomposition and localization always and at any cost, but just to the extent that the complexity of the system permits. Decomposition and localization remain in the picture to the extent that they are possible, practically interesting and explanatorily interesting. Consequently, the complex interactions of highly complex systems do not present indeed a big challenge. In this way, dynamical explanations do not oppose the mechanistic framework, nor do constitute an alternative to it. Instead, these are complementary endeavors, and dynamical explanations can be understood as a type of mechanistic explanations (Kaplan and Bechtel 2011, p. 438). Here these two kinds of explanation do not exclude each other, but on the contrary, support each other. At this fundamental level of analysis, and considering this particular issue, dynamicism then is not a threat to the centrality and popularity of mechanistic explanation as a base for understanding human cognitive processes and being perhaps the strongest fundamental theoretical framework in contemporary cognitive science and cognitive neuroscience.

5.1.3 Mechanistic-Dynamical Explanations and Strong Neuro-Cognitive Reduction

None of the arguments presented by dynamicists and mechanists, however, poses a problem for strong explanatory neuro-cognitive reduction, as long as it is correctly understood. As the literature on weak emergence shows (and also our analysis in Chap. 4), there is no incompatibility between complex dynamical systems characterizations and reductionism. Dynamicists engage in the same mischaracterization of reduction as that of the mechanists when they argue against reductive explanations. In their view, explanatory reductionism is decomposition and localization of individual parts and the construction of an explanation of the behavior of a given entire system just in terms of an individual part, or a set of parts and their respective operations. There is no clear and well-articulated account of reduction along such lines in the specialized literature. This would consist in an extremely controversial notion of scientific reduction—indeed a very easy target.

Our analysis, together with the specialized literature on reduction and emergence, shows to neo-mechanists and dynamicists that there is a more robust and strong type of reduction, constructed by Bickle (2003, 2006, 2008, 2012). This strong neuroscientific explanatory reduction has elements enough to stand against the attempts to argue for explanatory antireductionist positions by the proponents of these two frameworks.

Neo-mechanists and dynamicists, when debating about reduction and emergence, make the same mistake. They regard the operations of neural networks as something irreducible to and at a higher-level than molecular and cellular neural processes; thus, the processes of the latter cannot explain the processes of the former. However, activities of neural networks do not bring any substantial new causation to the world and can be completely reduced to their molecular and cellular components and their organization at that lower-level, which ultimately constitute them (cf. Chap. 4). For all these reasons, the claim that neo-mechanistic and dynamical explanations of cognitive processes are not reductionist in terms of ontology and epistemology is unsustainable.

5.2 TSHC Against MTHC

5.2.1 The Challenge to the Primacy of Neural Processes

The relatively recent research framework of the often so called 'situated cognition' is one of the most controversial theoretical achievements within the contemporary field of cognitive science. It has been described indeed as a new trend in cognitive science (Robbins and Aydede 2009, p. 3). Some authors claim that it belongs to a new generation of theories in cognitive science. Other authors go so far as to claim that this is the new science of the mind (Rowlands 2010).

Wilson and Foglia (2015, intro.) call the framework 'embodied cognition' and claim that cognition is embodied "when it is deeply dependent upon features of the physical body of an agent, that is, when aspects of the agent's body beyond the brain play a significant causal or physically constitutive role in cognitive processing". They further claim that this general theory encompasses a family of research programs in cognitive science and it has a commitment to critique or even replace traditional approaches of cognition in cognitive science. In this way, it stands as a serious alternative macro-theory to the investigation of cognitive phenomena as conceived by traditional theories of cognition in the field of cognitive science (cf. Clark 2003, 2008; Gibbs 2006; Hutchins 1995; Noë 2004, 2010; Shapiro 2004, 2011).

In the same line of argumentation, Shapiro claims that embodied cognition is a research program in its infancy that seeks "to replace, revise, or at least upset the reining [traditional] cognitivist conception of mind according to which cognitive processes involve computations over symbolic representations" (2012, p. 118). It is thus meant to be a new framework for studying cognition, often presented as an "alternative or challenger or 'next step in the evolution of' standard cognitive science" (Shapiro 2011, p. 1). Accordingly, the "common theme" here is that "traditional cognitive science" represented by a "reigning cognitivist orthodoxy" has "somewhere gone wrong, and that a fix requires attending more carefully to how an organism's body and interactions with the environment contribute to cognition" (Shapiro 2012, p. 142).

However, other authors do not see this theory with enthusiasm and point out to a variety of problems present in it. In their view, this framework does not provide a characterization of a well-defined and unified research program; it rather amounts to a more or less loose collection of a variety of ideas, scientific goals, philosophical considerations, empirical studies, theories and applications in cognitive science. Indeed, one finds in the literature a great amount of theoretical and terminological diversity, even to describe the main name of the framework, e.g. 'embodied cognition', 'embedded cognition', 'extended cognition', 'distributed cognition', 'enacted cognition', and 'situated cognition'. It is also not entirely clear what exactly the differences and similarities between all the ideas related to those terms are and if they indeed form a comprehensive more or less unified theoretical framework. Nevertheless, at least some main points of view are typically shared by authors working with this approach, and they can be analyzed and evaluated.

I will use the word 'situated' just because it seems the term more generally used by authors to refer to this framework (cf. Robbins and Aydede 2009; Walter 2014, chap. 5). It also appears to represent better the idea that cognition can be dependent on the body, environment, external devices, etc., since it does not restrict the concept as 'embodied' or 'embedded' cognition does. The central idea of situated human cognition, roughly, is that human cognition is situated in the sense that cognitive processes essentially depend upon our body and our embedding in and interaction with our natural, technological, and social environment. Accordingly, human cognition is not exclusively a neuronal process. The human brain plays, of course, a central role in human cognition, but the latter is also situated within a body that interacts with the environment, including other humans and technology—I call this theory the *Theory of Situated Human Cognition* (TSHC).

Authors working with this theory replaced 'abstract' problems of language comprehension and logical deduction by problems of active real-time interaction of physically embodied cognitive systems with their environment. They have observed, for example, that manipulating the environment is often an aid to problem solving. Consider, for instance, a neuro-prosthetic device that, similarly to a cochlea implant, restores a cognitive capacity that was originally implemented in the neural system. If such artificial substitutes can realize cognitive processes at all, it should not make a difference whether they are placed inside or just outside of the skull. Candidates for external constituents of cognitive processes are primarily artifacts: modern technologies, like pocket calculators, smartphones, computers, and classical tools, like pen and paper.

It is possible to distinguish at least two main theses here. The first one says that body and environment are important for the understanding of the occurrence of particular human cognitive processes. Cognitive processes are embedded in the sense that they depend upon processes in the environment. The second one says that human cognition is somehow constituted by brain-body-environment interaction (Dale et al. 2009, p. 739). It is thus a constitution claim: the environment is not a mere reservoir of cognitive resources that causally contribute to our cognitive performance; rather, many cognitive processes are currently realized not only in the neuronal tissue in our brain, but by hybrid processes comprising brain, body and technological and non-technological artifacts in the environment. Cognition thus is extended: cognitive processes are constituted in part by processes in the environment beyond the neural and bodily bounds of an organism. The claim is that the cognitive function itself, and the mechanism that realizes it, is partially in the environment: cognitive processes are sometimes constituted by hybrid processes that comprise both processes from within and processes from without the bounds of the organism. The idea here is that there is an intimate coupling between external resources and a cognitive agent, and the external resources eventually become integrated into the cognitive architecture of the agent performing functions that are usually performed by neuronal processes. This coupling would be enough for claiming that the external processes constitute the relevant cognitive processes.

The difference between these theses is that the weak one makes a claim of causal dependency; while the strong one makes a claim of constitution. Therefore, there is a (1) *weak thesis of TSHC*: human cognitive processes are constituted by brain processes only, however, bodily processes and environmental processes are extremely important as a causal influence for producing cognitive processes; they are not internal components of the system but merely external factors that affect the systems performance. And there is also a (2) *strong thesis of TSHC*: human cognitive processes can also be constituted by bodily and environmental processes. Such processes need to be, thus, considered as internal components of the entire system, not just as external factors that influence their behavior.

In order to better understand the strong thesis of situated cognition, which is an original one, I will systematically analyze a classic paper concerning the issue and present its main ideas. The paper is *The Extended Mind*, written by Andy Clark and David Chalmers and published in 1998.

In this paper, the authors ask the question concerning where does cognition (or mind) stop and the rest of the world begin. The issue here is about the *boundaries of cognition*. Some authors, as they say, advocate different positions. Some think that the boundaries of cognition are the skull (meaning that solely neural processes can realize cognitive processes); others think that bodily processes can also have an important role in realizing cognitive processes; others still go even farther and think that environmental processes can have such an indispensable role in realizing some cognitive processes. Clark and Chalmers aim to advocate a sort of thesis related to the third point of view, i.e. they advocate a view that emphasizes the active role of the environment in driving cognitive processes (1998, p. 7).

In order to support their view, the authors invite the readers to imagine a situation in which a person has the cognitive task of rotating two-dimensional geometric shapes in a computer screen in order to fit them into a particular space in the screen (like in the popular game 'Tetris'). In another scenario, the person can do this rotation not mentally but using a physical button, which physically rotate the geometric figure. In the last scenario, this rotation can be done by a neural implant. The authors argue then that all these cases are similar regarding cognition, and that the skull/skin boundary cannot be used as justification for claiming they are different, since the same cognitive function is been performed in all cases.

The authors argue further that human beings have a general tendency to lean heavily on environmental support in order to perform cognitive functions and/or improve their cognitive capacities. Many people, for instance, prefer to use pen and paper to perform mathematical calculations, or a pocket calculator, instead of doing it mentally, which requires typically more effort. Children, especially, tend to start getting used with mathematics by using their own fingers for making calculations. Equally, some writers might prefer to develop their new romance writing the new ideas and developing them at the same time, instead of thinking about everything just inside of their heads. The same applies to a scientist or a philosopher writing a new academic paper for publication. Even the use of books as repository of knowledge is a source for humans to remember that knowledge, instead of keeping it just in their minds. Currently and increasingly, people use personal computers and cell phones to store information and access information that they can use for their cognitive processes such as decision making or problem solving and to act based on this information. In this sense, the authors claim that the brain performs some cognitive operations, while external devices perform others (Clark and Chalmers 1998, p. 8).

The central idea for Clark and Chalmers is that if when a cognitive task is being performed a part of the environment "functions as a process which, *were it done in the head*, we would have no hesitation in recognizing as part of the cognitive process"; then that part of the environment "is […] part of the cognitive process" (1998, p. 8—highlights in the original). Basically, what is being advocated here is that bodily and environmental processes are component parts that constitute cognitive processes, not just neural processes play that constitutive role. It is not a matter of occasional causal interaction. As the authors point out: "The human organism is linked with an external entity… creating a coupled system that can be seen as a cognitive system in its own right"; if this external component is removed, "the system's behavioral competence will drop, just as it would if we removed part of its brain" (Clark and Chalmers 1998, pp. 8–9).

One could claim that this larger coupling of the situated-cognitive system is not reliable as the neuro-cognitive one, because the brain is always there when one needs it, but external media is not. In this way, it would be much easier to dismantle the situated-cognitive system, than the neuro-cognitive system. Clark and Chalmers answer that what is important is to have the component available whenever one needs it (1998, pp. 10–11). One could also always carry the component needed, e.g. when one carries a pocket calculator knowing that it will be necessary. Besides, the authors say, in the future, such systems could be plugged in the brain as additional-mechanisms (a mechanism for extra memory, for instance) in order to help us out. If the component is always there, when needed, then it is as reliable as the brain. Besides, occasionally the brain also loses its capacity, due to effects of sleep, intoxication and strong uncontrolled emotions.

In order to make their case, the authors also offer an interesting example related to beliefs and the memory capacity (Clark and Chalmers 1998, p. 12). They invite the reader to imagine the case of Inga, a person with all her neural and cognitive capacities working normally. Inga has a particular belief stored in her memory, which she can access when she wants. Another person, called Otto, however, suffers from Alzheimer's disease and, due to this problem, he relies on a notebook that he carries around with him everywhere he goes. When he learns new important information, he writes it in his notebook, and when he needs to remember this information, he accesses his notebook. Clark and Chalmers claim that Otto's notebook plays the same role normally played by the biological memory of Inga (1998, p. 12). As a result, information present in a system in the environment is being used for performing the cognitive function of remembering a certain belief: "In both cases the information is reliably there when needed, available to consciousness and available to guide action, in just the way we expect a belief to be" (Clark and Chalmers 1998, p. 13). The notebook represents to Otto what the biological memory represents to Inga. The final conclusion is that "there is nothing sacred about skull and skin. What makes some information count as a belief is the role it plays, and there is no reason why the relevant role can be played only from inside the body" (Clark and Chalmers 1998, p. 14).

In his textbook on philosophy of cognitive science, published in 2014, Andy Clark goes even farther than this. He claims that some cognitive processes arise given multiple factors. Some of the factors are bodily, some neural, some technological, and some social and cultural. Given this, the understanding of human cognition "may depend on a much broader focus than that to which cognitive science has become most accustomed". This focus includes "not just the body, brain, and the natural world, but the props and aids (pens, papers, PCs, institutions) in which biological brains learn, mature and operate" (Clark 2014, p. 167).

In the same book, Clark asks the readers to consider the example of the expert bartender, who, in a noisy and crowded environment, is asked to prepare multiple drinks, and performs this task with great accuracy. He is able to do that just because his skill involves an interaction with environmental components: the expert bartender selects different shaped glasses at the time of the ordering and use these cues to improve his capacity of memorize the sequence of drinks. The environment is thus used as a component in order to perform a cognitive function, simplifying the work of the biological brain. In this case, we have a part of the environment functioning as an extra-neural memory store (Clark 2014, p. 167). In conclusion, Clark (2014, p. 179) argues that, because of the process of human cognitive evolution, the human brain created a different way of dealing with cognitive technology, expanding in this way human reason. In his view, humans deploy nonbiological technology to complement basic biological dimensions of processing, and thus extended cognitive systems are created, which are very different from the brain alone (Clark 2014, p. 179).

What is being advocated here is clearly a larger (extended) cognitive mechanism, in which neural systems are just a component part; i.e. neural natural systems and other media such as a PC, or a cellphone, or pen and paper, are equally components performing an orchestrated operation.

To put it in a more formal way, for the sake of clarity and objectivity, what is being asserted by advocates of the strong thesis of TSHC is that the whole mechanism under consideration, *W*, is composed by a neural system, which is just one component part, C_1, performing its specific activities (functions), e.g. A_1, A_2, A_3. Another component part is, for instance, a PC or a cellphone, C_2, which performs its own operations, e.g. A_4, A_5, A_6. In this way, *W*'s structure is composed by C_1 and C_2 and *W*'s functions are composed by C_1's and C_2's functions, i.e. A_1, A_2, A_3, A_4, A_5, A_6. In this picture, the external factors that might causally influence the structure, E_s, and function, E_f, of *W* are something else. While the mechanists would say that the mechanism in question is composed by C_1 (and its subcomponents) solely, the advocates of situated cognition claim it is composed by C_1 and C_2 (and their subcomponents). This is precisely where they differ. For the advocates of the strong thesis of TSHC, there is a strict pattern of continuous, reciprocal and highly interactive causal influence between C_1 and C_2 so that both are united and count as components of the same entire system. Were such strict interaction not present, C_2 would have to be considered just an external causal factor that exercises a relatively independent influence on the system's performance from outside.

Therefore, in order to provide an answer to defenders of TSHC, formulated in this strong way, the mechanists need to argue that C_2 cannot or should not be considered equally a component in *W*; at best, they can say, it is just an external factor that affects *W* structurally or functionally, but not a component part of *W*. They need to show that the work being done by C_2 is actually being done by C_1, i.e. that the functions being attributed to extended components are in fact functions being fully performed by some neural system. And they need, of course, to provide good reasons and good arguments in order to support this point of view.

5.2.2 The Mechanists' Reply: The Emphasis on the Interaction with External Factors

Unfortunately, in the relevant specialized literature it is still difficult to find publications that deal in a more systematic and detailed way with this controversy between neo-mechanists and contemporary advocates of situated cognition. However, there is at least one paper where William Bechtel himself addresses the issue in a more detailed fashion. The paper is called *Explanation: mechanism, modularity, and situated cognition.* It was published in 2009 as a chapter in the *Cambridge Handbook of Situated Cognition.* In what follows, I make a methodical analysis of this work.

In this paper, Bechtel starts with the recognition that: "The situated cognition movement has emerged in recent decades [...] largely in reaction to an approach to explaining cognition that tended to ignore the context in which cognitive activities typically occur" (2009, p. 155). He claims that the work of Jerry Fodor (1935–2017) is a representative of this approach, since, for Fodor, "only representational states within the mind are viewed as playing causal roles in producing cognitive activity" (Bechtel 2009, p. 155). The central thesis of the situated cognition approach is identified by Bechtel as insisting that "an agent's cognitive activities are inherently embedded and supported by dynamic interactions with the agent's body and features of its environment" (2009, p. 155). This is the 'weak' thesis that we have identified in the previous section. And, as regards this weak thesis, Bechtel seems to see no problem or incompatibility with his mechanistic theory. On the contrary, in another work he manifests actually some sympathy towards it. He claims that the environment and social situations are important for psychological explanations and that increasingly psychologists are interested in explanations concerned with embodied and situated contexts (Bechtel and Wright 2009, p. 127). However, Bechtel also identifies the strong thesis in the approach of situated cognition:

> Sometimes advocates of a situated approach to cognition present their position in an extreme manner that sets the situated approach in opposition to attempts in cognitive science and cognitive neuroscience to understand the mechanisms within the mind/brain that underlie cognitive performance. (Bechtel 2009, p. 155)

This means that some advocates of the situated approach hold that cognitive processes do not happen just in the brain, but bodily and environmental factors can be also considered as constituting (being also a part) of cognitive processes.[2] These authors "reject drawing a sharp distinction between the mind/brain and the rest of the body and environment in which the mind operates" (Bechtel 2009, p. 158). In this sense, the thesis clearly opposes the thesis that solely neural mechanisms constitute cognitive processes, which underlie the vast majority of work in cognitive science and cognitive neuroscience.

There are, therefore, two theses that collide. Firstly, the s*trong thesis of TSHC*: cognitive processes are constituted by brain activity, but also, at least sometimes, by bodily processes and environmental processes. It is, therefore, not just a matter of causal influence, or causal interaction. And, secondly, *the mechanists' thesis*: cognitive processes are constituted solely by brain activity, even if bodily processes and environmental processes play an important causal role for producing them.

Accordingly, Bechtel argues that "an appropriate understanding of the situated cognition does not require the denial that the proper locus of control [...] for cognitive activity is the mind/brain" (2009, pp. 155–156). This means that for cognitive phenomena "it is appropriate to treat the mind/brain as the locus of the responsible mechanism and to emphasize the boundary between the mind/brain and the rest of the body and between the cognitive agent and its environment" (Bechtel 2009,

[2] Bechtel actually cites at this point the work of Clark and Chalmers (1998) in his paper, but no quotation is offered to better characterize the position, much less any particular analysis of any of its content.

p. 155–156). However, he also recognizes that this strategy is not appropriate for a certain kind of phenomena, i.e. there are cases in which the agent is so strictly related to external factors that the responsible mechanism includes one or more cognitive agents and their environment; but such phenomena are, according to him, typically social phenomena, not behavioral or psychological (Bechtel 2009, pp. 155–156). There is no cognitive extension here, then. At any rate, Bechtel claims that some explanatory principles can be used to determine more precisely when it is adequate to identify the mind/brain as the locus of control and when it is appropriate to identify a larger system as responsible (2009, p. 156). He contends, nevertheless, that "for most explanatory challenges addressed by cognitive science, the mind/brain is the appropriate locus of control even for activities that depend critically on how the agent is situated in an environment" (Bechtel 2009, p. 156).

In the paper under analysis, therefore, Bechtel attempts to reconcile "the cognitive science project of identifying and describing the operations of mechanisms inside the head with the claims that cognition is situated" (2009, p. 156). In order to do that, he argues that biological mechanisms (including cognitive mechanisms) "are bounded systems", on the one hand, but at the same time they "are selectively open to their environment" and they "often interact with and depend on their environment in giving rise to the phenomenon for which they are responsible" (Bechtel 2009, p. 156). This means that such mechanisms can be considered organisms that interact and are dependent on their environments, but at the same time are distinct in respect to this environment.

Equally, Bechtel argues against the thesis of strong modularity. As he explains, advocates of strong modularity identify the responsible place for cognitive activities at a "finer level", i.e. "not with the whole mind/brain but with a module within it" (2009, p. 156). The *thesis of strong modularity* holds that cognitive processes are constituted not by the whole mind/brain, but by a unique module inside the brain. However, this thesis, in Bechtel's view, is wrong, since it "fails to consider that explaining the mind/brain's performance of a cognitive task involves decomposing it into component operations, each of which contributes differentially to the performance of the task" (2009, p. 156). Besides, one subsystem of a cognitive system is not enough to perform most cognitive tasks, since they require the orchestrated contribution of many components of this system (Bechtel 2009, p. 156).

This strong thesis of modularity is thus rejected because it fails to "recognize the diverse components involved in performing a cognitive task", on the one hand (Bechtel 2009, p. 156). The strong thesis of situated cognition, on the other hand, is rejected because it maintains that "many cognitive activities involve components outside the agent itself" (Bechtel 2009, p. 156). Basically, strong modularity requires too less, i.e. a sub-mechanism smaller than the whole mechanism to account for the activity of the whole mechanism, while strong situated cognition requires too much, i.e. it states that we need more components than those that we actually need. The mechanistic thesis, in Bechtel's view, is thus the correct one, since it avoids both these problems.

The mechanistic thesis recognizes that: (1) all the components in a whole biological mechanism need to be considered in the explanation of a function performed by the whole mechanism, not just a sub-system of this mechanism (note that this resembles very much the mischaracterization of the strong neuroscientific reductionist thesis); (2) components in a biological mechanism are highly interactive with other mechanisms and this is why they are important for the production of the phenomenon; (3) biological mechanisms have an identity of their own and cannot be confused with processes in their environment, even if these mechanisms are highly affected by their environment, which influences the production of their processes.

In order to argue against the strong modularity thesis, Bechtel offers firstly a simple account of a 'module', based, as he says, in the work of Fodor and evolutionary psychologists. A module, according to his characterization, would be then a particular mechanism responsible for a specific type of processing whose most important feature is the use of information just from its own processes, not from outside; it is informationally closed and does not interact informationally with other modules (Bechtel 2009, p. 157). Advocates of modules, he says, do not work with decomposition of a whole system into component parts and operations; a module is thus a domain of performance—the model performs one of the tasks performed by the whole system (Bechtel 2009, p. 160). Such characterization is used by Bechtel to claim that his conception of components of a mechanism is very different from the conception of modules (2009, p. 160).

The reason is that in mechanistic explanations "investigators often learn that other parts are involved in producing the phenomenon and that the part in question only performs one of the required operations", this results in "a theory of an integrated system responsible for the phenomenon, not a single component" (Bechtel 2009, p. 160). To put the point in another way: "The fact that the operations that parts of a mechanism are different from the phenomenon exhibited by the whole mechanism and individually do not realize the phenomenon makes the working parts of a mechanism different from domain-specific modules"; e.g. the "cardiovascular system of an organism [...] consists of distributed parts that perform different operations"—"neither such mechanisms themselves nor their parts are encapsulated [informationally closed] from each other in the manner of Fodorian modules" (Bechtel 2009, p. 160).

Moreover, Bechtel believes that if components in a system are informationally closed, this system can be fully decomposed. However, this does not happen in biological systems where even in relatively simple organisms there is a huge number of interactions, which quickly reduces the decomposability of the overall system (Bechtel 2009, p. 162). In fact, biological systems are highly integrated despite the number of different operations realized by different organs and cell types (Bechtel 2009, p. 167). Therefore, decomposability is merely a heuristic principle of mechanistic explanations of biological phenomenon and can be achieved just as much as possible (cf. this chapter, Sect. 5.1.2). Subsystems can be partially methodologically isolated so that the interaction of the entire system is made intelligible in terms of the parts, but it is also important to recognize that components interact in very complex ways, which need to be better understood with the development of the mechanistic explanation.

In order to argue, in turn, against the situated cognition approach, Bechtel takes a particular direction. His aim is to show that biological organisms need to be distinguished from their environments and that some mechanisms within biological organisms need to be also distinguished from each other. The reason for this is that living systems require metabolic processes that use energy and transform it in particular activity. These systems rely on biological membranes to separate themselves, and component systems inside them, from their environment. Moreover, they admit inside themselves selectively substances necessary for their own maintenance and development, while they remove from themselves substances that are no longer needed (Bechtel 2009, p. 163). They are, therefore, autonomous biological systems which actively engage to maintain themselves, and each of the internal mechanisms that constitute them are important for their survival (Bechtel 2009, p. 163). Moreover, complex organisms are composed of different organ systems, while each individual organ performs different operations that are required by the whole organism (Bechtel 2009, p. 165). Such internal systems (as components) are not informationally closed to other components of the system, but open in appropriate ways to them (Bechtel 2009, p. 166).

Thus, "the mind/brain", as "a part of the organism", is "differentiated from the world in which it operates but is nonetheless highly connected to that world": there is, thus, "both isolation from the environment as the organism maintains its own identity and engagement with it" (Bechtel 2009, p. 166). Accordingly, although such systems are not isolated and "totally independent of their environment", "it is appropriate to construct them as different from their environments" (Bechtel 2009, p. 164). Ultimately, it is "still the cognitive agent that is performing these activities in pursuit of its ends. It is the cognitive agent that has an interest in performing the task and in recruiting components of its environment to enable such performance" (Bechtel 2009, pp. 166–167). In this sense, "the cognitive agent is like autonomous biological systems that perform operations in their environment so as to secure matter and energy needed to build and repair themselves and dispose of wastes that are toxic to them" (Bechtel 2009, pp. 166–167). Accordingly, there is no need to affirm that the mind extends "out into the world", nor to deny the "differentiation of the mind/brain from the rest of the organism and the external world" (Bechtel 2009, p. 165). The mind/brain is composed by many highly integrated neural areas that perform different computations; there is no informationally closed system in this mechanism (Bechtel 2009, p. 167). However, it is important to understand that such mechanism has boundaries, even though it depends on the environment and interacts with it.

5.2.3 The Mechanists' Reply Under Analysis

In my view, Bechtel's account contains mischaracterizations not just of the strong thesis of TSHC but also of the thesis of modularity. Virtually, he wants us to believe that there are authors claiming something like 'a certain specific part of a watch keeps time, not the entire watch composed by all its parts, their operations and interactions'; or 'a certain particular component part of the heart is capable of pumping blood to the entire circulatory system, not the whole heart performs this function'. But no author is claiming something of this sort. Bechtel does not offer a single concrete example of an author defending seriously a point of view such as this one. The reason is that it is trivially not true. This characterization is quite similar to the characterization of neuro-cognitive reductionism by some influential neo-mechanists (cf. Chap. 4). Both these characterizations are trivially not true and, therefore, not seriously defended by any author in the literature. In this way, Bechtel is interpreting 'modularity' and 'neuro-cognitive reductionism' in a very similar manner, and he is giving a very partial interpretation of both of them. Both interpretations can be thus forcefully challenged.

Furthermore, Bechtel's arguments against the *strong thesis of TSHC* are beside the point at issue. The information that cells have membranes which differentiate them from their environments and that cells interact with their environments in order to acquire energy and eliminate wasted products is something well known. The same is true concerning the information that we have in our organism different organs and systems that interact and make us a whole biological system able to survive in our environments. But Clark and Chalmers are not concerned with biological cells and their membranes, nor are they saying that there is no physio-biological difference between the brain and the environment surrounding it. Their point is about *cognitive processes* that can be extended into the body and the environment. The issue is not about a biological extension, but a cognitive one. And if one assumes that cognitive processes are neural processes in order to argue that the brain must be differentiated from the environment, one is already assuming what is precisely the matter in dispute.

Let us take the example of Otto and his notebook for the cognitive capacity of memory consolidation. What Clark and Chalmers claim is that there is no difference if the process of memory consolidation is achieved in the brain of Otto by memorizing it (with all the neural aspects involved) or by his writing in his notebook, since in the end, the cognitive task of remembering some particular event will be performed basically in the same way. In this case, Bechtel and the neo-mechanists should show that the brain performs a special procedure that is unique, and that one does not find this in the process that happens between Otto and his notebook. What needs to be clear is this: what exactly is the cognitive difference between memorizing some information in the brain and writing it on a digital notebook, or a smartphone? Why memorizing using a digital notebook does not qualify as genuine cognition, while the processes done by the particular relevant neural mechanism qualify? Where precisely lies the difference? In order to challenge substantially the thesis of Clark and Chalmers, the neo-mechanist needs to show that there is a significant and relevant difference between 'notebook-memory', or 'cellphone-memory', and 'brain-memory'. Based on this clarification, an argument can be constructed in order to show that cognitive processes do not happen outside of the neural system. Perhaps neo-mechanists could appeal to some sort of mental representation mechanism that is available just in the case of neural systems, or some sort of particular neural computation or neural information processes that are idiosyncratic to the neural system giving rise to the particular cognitive function under consideration. But such argument is not presented by neo-mechanists—perhaps because no one of them has yet a clear and reasonably articulated view of what exactly this special procedure would be. When one goes in that direction, one starts confronting some of the problems related to the mind-body relationship, and the issue of the nature of human cognition. This is why the discussions become so controversial. An outlined account of the nature of human cognition and how it is particularly related to the brain must be presented so that the boundaries of human cognition can be established.

The argument advanced by Bechtel in his paper shows that there is interaction between cognitive processes and brain processes with the environment, and that the environment influences these processes. However, the fact that cognition is influenced by and interacts with external objects and processes is true and trivial. There is a vast amount of studies showing this in particular cases—consider, for instance, traditional studies in psychology related to perception and sensation. But, in fact, the general idea is common sense. For example, in books, papers and other written materials it is often possible to find useful information that can be considered in order to make decisions; this is why so many people read them (Adams and Aizawa 2009, p. 79). The issue concerning the strong thesis of situated cognition, however, is about the constitution of a given mechanism responsible for performing a particular cognitive function (i.e. the thesis that all or some cognitive processes extend to and are constituted by objects and processes beyond the brain), and not merely external causal occasional influence or interaction. No one in this debate is asserting that an organism does not interact frequently with its natural environment in many different significant ways. Bechtel, nevertheless, makes the case for the external causal influence and interaction, i.e. the trivial thesis—precisely what is beside the problematic point being debated.

Still, there are indeed many ways of formulating a criticism to the strong thesis of TSHC. Adams and Aizawa, for instance, consider the hypothesis as "outrageous" (2010, p. vii) and "crazy" (2010, p. viii) and argue forcefully that the ideas of Clark, Chalmers and other proponents of the framework are mistaken. In their critique, Adams and Aizawa attempt to outline a positive view on what genuinely counts as cognition, considering basic presuppositions of contemporary cognitive psychology. They maintain that "cognitive processes use particular kinds of mechanisms that operate on specific kinds of mental representations" (Adams and Aizawa 2010, p. viii). Furthermore, they claim that "there is something distinctive about the brain", certain "natural kinds of processes […] happen to occur only within the brain"; and such processes "differ from neurophysiological processes insofar as they consist of […] causal operations on non-derived representations (representations whose content does not depend on other previously existing content"; in addition, such processes also differ "from typical processes that extend into the world from brains and from processes found in typical machines" (Adams and Aizawa 2009, p. 80). In other words, there is a "distinct type of information processing capacities of the brain" that account for the nature of cognition (Adams and Aizawa 2009, p. 80).

With this view, the authors can distinguish between cognitive processes and non-cognitive processes. For them, genuine cognitive processes are thus brain bound, rather than spread over brain, body and environment. The authors state that "our cognitive faculties, restricted to the confines of our brains, can be aided in any manner of ways, by cleverly designed non-cognitive tools", such as calculators, books, slide rules, computers, microscopes, telescopes, etc. (Adams and Aizawa 2001, p. 44). However, the processes related to such tools *cannot* be considered cognitive processes, despite of their interaction with our cognitive faculties. Consequently, in their view, as a matter of contemporary contingent empirical fact "the mind is still in the head" (Adams and Aizawa 2009, p. 78).

They accept that there are many cases in which there is extensive interactions between cognitive processes and environmental objects, as in the case when pencil and paper is used to compute sums of large numbers. But, in their view, these interactions do not provide good reason to think that the processes related in this case to the paper and the pencil are cognitive themselves. For a cognitive process can interact with an X process and this does not mean that the X process is itself also a (part of a) cognitive process. This is the "coupling-constitution fallacy" and it involves a "confusion of coupling relations with constitutive relations" (Adams and Aizawa 2009, p. 78).

Moreover, they argue that the advocates of situated cognition do not provide an account of what really cognitive processes are, which remains lacking in the framework. The insufficient attention to what makes a process cognitive ('the mark of the cognitive') and/or the advance of a loose conception of cognitive makes the hypothesis to appear more plausible than it really is (Adams and Aizawa 2010, p. ix). In their view, once the knowledge about cognitive processes available in contemporary cognitive science is well articulated and the basic necessary conditions for a process to be considered genuinely cognitive are established, one notes that the theory of situated cognition is implausible.

Finally, giving the vast variety of possible situated-cognitive processes and systems permitted by the theory, one could argue that the concept of cognitive processes and thus the object of cognitive science loses completely its limits and consequently cannot be used meaningfully as a topic of serious scientific explanation. Standard cognitive science at least "helps us define a field of scientific investigation. It offers us a relatively well-defined subject [...] to study. It is the study of some, but only some, of the ways in which some organisms transform information, ways it might be possible to recreate in computers or robots" (Adams and Aizawa 2010, p. 580). Thus, it helps "to demarcate the regions of the world that *contain* processes versus those that do not" (Adams and Aizawa 2010, p. 580—highlight in the original). Advocates of situated cognition "have not provided a plausible successor to the idea that cognitive science is about some of the ways in which some organisms manipulate some representations"; hence, they "do not have a plausible account of what cognitive science is a science of" (Adams and Aizawa 2010, p. 581).

A detailed criticism of that kind is what Bechtel and other influential neo-mechanists should provide in order to argue for the superiority of MTHC in cognitive science and to unify and integrate the field, as they intend to. Bechtel, however, does not develop any criticism of that sort or even mentions any of this. Instead, he argues for a point that is not crucial to the discussion. Other very influential neo-mechanists appear simply to ignore the framework of situated cognition in cognitive science, offering no account on how exactly it relates to MTHC (cf. Boone and Piccinini 2015).[3] Therefore, the strong thesis of TSHC, standing as it does currently—without being seriously and effectively challenged by influential neo-mechanists— remains as an alternative framework in cognitive science.

And it has at least two consequences for the mechanistic point of view concerning the study and explanation of human cognitive phenomena in cognitive science. The first consequence is that, in order to understand the situated-cognitive biological systems of humans (with heterogeneous components), the brain cannot be isolated from the other components of the system; otherwise the causal interactions between the internal components of the mechanism will be lost. Accordingly, the focus cannot be primarily on brains or neural systems (as subcomponents); rather, the entire system will have to be considered. However, the entire system can be too large to be scientifically treated given our current scientific capabilities. A possible solution is to try to consider the simplest mechanisms in which all the interactions (neural, bodily, environmental, and technological) are present and start building theories of their activity from there (Clark 2014, p. 184). The second consequence is that the scope of cognitive science is considerably amplified with the situated theory. Cognitive science would be thus a larger scientific enterprise which includes multidisciplinary work in neuroscience and physiology, sociology and cultural studies and studies on technology and artificial cognitive systems in equal measure (Clark 2014, p. 185). Given that neural phenomena do not play any major role, there is no compelling reason to treat cognitive neuroscience as a more important science than other sciences that belong to cognitive science, as the mechanists clearly do. Much less to equate cognitive science to cognitive neuroscience or, worse, to treat cognitive science, or cognitive psychology, as a science that belongs to neuroscience, as some mechanists do (e.g. Craver 2007).

With this analysis I am not establishing (and I am not aiming at establishing) that the strong thesis of TSHC is true. It is certainly not my purpose to defend any form of TSHC here. Nor I am establishing that there certainly is an incompatibility between MTHC and the strong thesis of TSHC. What I am establishing is that Bechtel and other neo-mechanists still cannot show that the strong thesis of situated cognition is wrong; and if they think it is right, they have not shown yet how it could be made compatible with their mechanistic framework. The strong thesis of TSHC remains thus a challenge and an alternative to MTHC, since currently both frameworks present incompatible positions.

5.3 Final Remarks

In this chapter, I argued that MTHC can deal with the criticism of the advocates of DSTHC with respect to the difficulties concerning decomposition and localization. The mechanistic framework can simply recognize that at higher-levels of

[3]But see Milkowski et al. (2018) for a discussion.

complexity the organization of the whole system becomes too complex, to the extent that highly specialized decomposition and localization are not possible for such particular systems. This, however, does not bring any problem for a causal mechanistic explanation in the end, based on the dynamics of these systems and their complex causal relations, since the mechanists also emphasize organization and do not put any especial requirement on what kind of organization that must be. At the same time, even in highly dynamical complex systems, some degree of decomposition and localization can still be useful in order to help dealing with the great amount of complexity and favor more dynamical analysis. Therefore, there is indeed no conflict, but rather compatibility between the two frameworks.

Concerning TSHC, however, the issues are more complex. The analysis carried out in this chapter shows that MTHC and strong TSHC are not yet compatible, since their advocates defend opposing views and, so far, no plausible proposal of making them compatible was offered. Consequently, strong TSHC remains as an alternative to the neo-mechanistic theory. Yet, concerning TSHC, many issues still remain controversial. Can extracranial neuronal prosthetics processes, cell-phones processes or notebooks processes be also considered constituents of cognitive processes? Or are they merely causal contributors to cognitive processes? The differences and boundaries between the constituents of a cognitive system and those processes that causally contribute to its functioning need to be clarified. Regarding human cognitive systems it is not entirely clear where to establish this boundary and why to establish it exactly there. It is very important, thus, to better clarify this issue of 'the mark of the cognitive', and develop a standard and well accepted theory of human cognition that could tell us what human cognition substantially is and is not—otherwise the debates about situated cognition are likely to remain controversial.

References

Adams, F., Aizawa, K.: The bounds of cognition. Philos. Psychol. **14**(1), 43–64 (2001)

Adams, F., Aizawa, K.: Why the mind is still in the head. In: Robbins, P., Aydede, M. (eds.) The Cambridge Handbook of Situated Cognition, pp. 78–95. Cambridge University Press, Cambridge (2009)

Adams, F., Aizawa, K.: The Bounds of Cognition. Wiley-Blackwell, Malden, MA (2010)

Bechtel, W.: Representations and cognitive explanations: Assessing the dynamicist's challenge in cognitive science. Cognitive Science **22**(3), 295–318 (1998)

Bechtel, W.: The compatibility of complex systems and reduction: A case analysis of memory research. Mind. Mach. **11**, 483–502 (2001)

Bechtel, W.: Mechanism, modularity, and situated cognition. In: Robbins, Philip, Aydede, Murat (eds.) The Cambridge Handbook of Situated Cognition, pp. 155–170. Cambridge University Press, Cambridge (2009)

Bechtel, W., Wright, C.D.: What is psychological explanation? In: Symons, John, Calvo, Paco (eds.) The Routledge Companion to Philosophy of Psychology, pp. 113–130. Routledge and Taylor & Francis Group, New York (2009)

Bickle, J.: Philosophy and Neuroscience: A Ruthlessly Reductive Account. Kluwer Academic Press, Norwell, MA (2003)

Bickle, J.: Reducing mind to molecular pathways: Explicating the reductionism implicit in current cellular and molecular neuroscience. Synthese **151**, 411–434 (2006)
Bickle, J.: Real reduction in real neuroscience: Metascience, not philosophy of science (and certainly not metaphysics!). In: Hohwy, J., Kallestrup, J. (eds.) Being Reduced: New Essays on Reduction, Explanation, and Causation, pp. 34–51. Oxford University Press, Oxford (2008)
Bickle, J.: A brief history of neuroscience's actual influences on mind–brain reductionism. In: Gozzano, S., Hill, C.S. (eds.) New Perspectives on Type Identity: The Mental and the Physical, pp. 88–110. Cambridge University Press, Cambridge (2012)
Boone, W., Piccinini, G.: Cognitive neuroscience revolution. Synthese (2015). https://doi.org/10.1007/s11229-015-0783-4
Chemero, A.: Radical embodied cognitive science. MIT Press, Cambridge, MA (2009)
Chemero, A., Silberstein, M.: After the philosophy of mind: replacing scholasticism with science. Philos. Sci. **75**(1), 1–27 (2008)
Clark, A.: Natural-Born Cyborgs. Oxford University Press, Oxford (2003)
Clark, A.: Supersizing the Mind. Oxford University Press, Oxford (2008)
Clark, A.: Mindware: An Introduction to the Philosophy of Cognitive Science, 2nd edn. Oxford University Press, Oxford (2014)
Clark, A., Chalmer, D.: The extended mind. Analysis **58**, 7–19 (1998)
Craver, C.F.: Explaining the Brain: Mechanisms and the Mosaic Unity of Neuroscience. Oxford University Press, Oxford (2007)
Dale, R., Dietrich, E., Chemero, A.: Explanatory pluralism in cognitive science. Cogn. Sci. **33**, 739–742 (2009)
Felleman, D.J., Van Essen, D.C.: Distributed hierarchical processing in the primate cerebral cortex. Cereb. Cortex **1**, 1–47 (1991)
Gibbs, R.W.: Embodiment and Cognitive Science. Cambridge University Press, Cambridge (2006)
Hutchins, E.: Cognition in the Wild. MIT Press, Cambridge, MA (1995)
Kaplan, D.M., Bechtel, W.: Dynamical models: An alternative or complement to mechanistic explanations? Top. Cogn. Sci. **3**, 438–444 (2011)
Milkowski, M. et al.: From wide cognition to mechanisms: A silent revolution. Front. Psychol. **9** (2018). https://doi.org/10.3389/fpsyg.2018.02393
Noë, A.: Action in Perception. MIT Press, Cambridge, MA (2004)
Noë, A.: Out of Our Heads. Hill & Wang, New York (2010)
Robbins, P., Aydede, M.: A short primer on situated cognition. In: Robbins, P., Aydede, M. (eds.) The Cambridge Handbook of Situated Cognition, pp. 3–10. Cambridge University Press, Cambridge (2009)
Rowlands, M.: The New Science of the Mind: From Extended Mind to Embodied Phenomenology. MIT Press, Cambridge, MA (2010)
Shapiro, L.: The Mind Incarnate. The MIT Press, Cambridge, MA (2004)
Shapiro, L.A.: Embodied Cognition. Routledge, New York (2011)
Shapiro, L.A.: Embodied cognition. In: Margolis, E., Samuels, R., Stich, S. (eds.) The Oxford Handbook of Philosophy of Cognitive Science, pp. 118–146. Oxford University Press, New York (2012)
Stepp, N., Chemero, A., Turvey, M.T.: Philosophy for the rest of cognitive science. Top. Cogn. Sci. **3**, 425–437 (2011)
Van Gelder, T.: What might cognition be, if not computation? J. Philos. **91**, 345–381 (1995)
Van Gelder, T.: The dynamical hypothesis in cognitive science. Behav. Brain Sci. **21**, 615–628 (1998)
Walter, S.: Kognition. Reclam, Stuttgart (2014)
Wilson, R.A., Foglia, L.: Embodied cognition. The Stanford Encyclopedia of Philosophy (Spring 2017 Edition). In: Zalta, E.N. (ed.) (2015). https://plato.stanford.edu/archives/spr2017/entries/embodied-cognition/

Chapter 6
Computational Theory of Human Cognition and Belief-Desire Theory of Human Cognition

6.1 CTHC and BDTHC Against MTHC

6.1.1 Introduction to the Challenges

There are still two other alternative theories to MTHC in the broad field of cognitive science. The first one is the *Computational Theory of Human Cognition* (CTHC),[1] which arises together with the classical research program in cognitive science in the decades of 1960 and 1970 and is more particularly related, for example, to the subfield of computational cognitive science. This theory assumes that human cognition is literally a kind of computation. Moreover, the majority of researchers working in accordance with this theory is connected usually to institutions like the *Cognitive Science Society*, for example; and they publish more frequently in journals such as *Cognitive Science*, *Topics in Cognitive Science*, or *Cognition*, for example.

The second alternative theory is the *Belief-Desire Theory of Human Cognition* (BDTHC), which has been advocated traditionally in the field of psychology and is more particularly related, for example, to the subfields of developmental psychology, social psychology, and educational psychology.[2] This theory is not committed to the thesis that human cognition is a kind of computation. This is the major distinction between CTHC and BDTHC, which most significantly separates the two theoretical frameworks. BDTHC accepts, however, that human cognitive capacities can be generally understood in terms of functions and the relations between them—these functions can be also related to the environment and to the behavior connected to them. Furthermore, classical concepts as used by traditional scientific psychology (e.g. the concepts of 'belief', 'desire', 'intention', etc.) are indispensable for this theory. The majority of researchers working in accordance with this theory is connected usually to institutions like the *American Psychological Association*, for example; and they

[1]cf. Rescorla (2020).

[2]On these characterizations, I am following Cummins (2000), even though I do not follow exactly his terminology.

D. A. Leite, *The Twenty-First Century Mechanistic Theory of Human Cognition*, Cognitive Systems Monographs 41, https://doi.org/10.1007/978-3-030-63680-7_6

publish frequently in journals such as *Psychological Bulletin, Psychological Review*, and *American Psychologist*, for example. It is also important to distinguish BDTHC from the so-called 'folk psychology' taken in the sense of the everyday human ability to understand and explain their own mental activity and the mental activity of others. As is well known, folk psychology has many shortcomings and is not generally considered a scientific theory of psychology (cf. Weiskopf and Adams 2015, p. 2). Similarly, folk physics (roughly, everyday explanations of natural events by common people) is not considered a scientific theory in physics.

CTHC and BDTHC can be classified as clear forms of 'cognitivism' (or 'mentalism'), taken in the sense of a doctrine which asserts that 'cognition' or 'mind' is real (not an illusion) and deserves scientific attention. Cognition here, then, is not the same as 'behavior'; it is 'something else'. Even though cognitive capacities cannot be directly observed, cognitive science assumes they can be inferred using scientific experimentation and scientific tools. They are generally considered to be 'intervening variables' that are indispensable for explaining particular kinds of behavior presented by humans and other animals. These two frameworks, therefore, are committed to a view of cognitive phenomena and explanation in cognitive science as something at least partially independent of neural phenomena and neuroscientific explanation. Consequently, they present a challenge to the attempt of the neo-mechanists to integrate psychological explanation and phenomena to neural explanations and phenomena, which is indeed one of the most difficult and controversial tasks for theoretical cognitive science in the twenty-first century.

MTHC proposes a kind of integration that includes repelling the autonomy of psychological explanations, on the one hand, while, on the other, it includes the attempt to retain certain particular notions of cognitive information processing, computation and representation as fundamental pillars of the general framework. Given this, one can ask: are explanations in cognitive science always constructed in terms of biological mechanisms? Are all psychological explanations provided by CTHC and BDTHC mechanistic explanations? Can all human cognitive capacities be simply considered physio-biological mechanistic capacities? In this chapter, I use some of the most important well-known features of CTHC and BDTHC to inform my analysis of these central topics, meticulously discuss each of these questions and provide some answers to them.

6.1.2 Challenge I: The Nature of Explanations in Cognitive Science

(A) *Laws in Psychological Explanations*

One should not simply assume that explanations of phenomena will be constructed according to the same (or even a similar) structure across all the sciences. The mechanistic theory states that scientific explanations in some scientific fields are not constructed in terms of laws, but rather in terms of mechanisms (Chap. 2, Sect. 2.1.3).

Nevertheless, the scope of the theory is not yet clear (cf. Rosenberg 2015). Sometimes the scope is stated in a vague form: many scientific works are directed to the search for mechanisms (Craver and Tabery 2015, Sect. 1). This does not tell us whether the framework is to be applied just to all the biological sciences, or also to all the so-called special sciences (e.g. history, economy, sociology, political science, law, computer science, etc.), or even to chemistry and physics (cf. Craver and Tabery 2015, Sect. 2.6). Evidently, the mechanistic framework needs to be better elaborated also regarding the issue of its scope of application to the scientific fields. However, it is clear that to biology, neuroscience and cognitive science it is applied. As Bechtel points out, "the deductive-nomological model runs into problems in disciplines that possess few laws but many purported explanations—for instance, biology [...], psychology and other cognitive sciences" (2009a, p. 552). My focus, at any rate, is in cognitive science: particularly here in the central claim made by influential neo-mechanists according to which there are (almost) no laws in cognitive science.

This is a highly controversial issue, since one can argue, on the contrary, that laws are present in cognitive science. One example is *Fechner's law* established in the nineteenth century, in the field of psychophysics, by the German experimental psychologist Gustav Fechner (1801–1887). The general idea is to understand how the properties of sensations depend on and vary with the properties of physical stimulus that produce them (Weiskopf and Adams 2015, p. 7). Phenomena such as light, sound waves and temperature interact with sensory receptors, giving rise to sensations. Fechner's work assumes that there is a systematic and regular relationship, rather than a random one, between those external phenomena and the internal psychological sensations they produce in the subjects (Fechner 1889). To better understand this relationship, it was necessary to empirically measure the sensations, quantify the covariation between sensations and the stimulus conditions, and explain this covariation presenting a law. Fechner controlled the increase in the intensity of the stimulus and could measure the "intervals at which a detectable change in sensation occurred against the stimulus that caused that change" (Weiskopf and Adams 2015, p. 7). Fechner's law can be expressed by a simple equation: $S = k \log (I)$, where "S is the perceived magnitude of the sensation (e.g. the brightness of a light or the loudness of a sound), I is the intensity of the physical stimulus, and k is an empirically determined constant" (Weiskopf and Adams 2015, p. 7—highlights in the original). According to the law, increases in the intensity of the stimulus will correspond to increases in the strength of sensations. Other laws have also been established in the field and are mentioned in the relevant literature with some frequency. For instance, *Weber's law* and *Steven's power law* concerning sensations, and the *Matching law* and *Thorndike's law of effect*, concerning learning and behavioral conditioning. Therefore, it appears that laws are present in cognitive science.

However, neo-mechanists frequently point out that the empirical generalizations in cognitive science are fragile: they have many exceptions and are often relative to a given population of subjects, in a given time and in a given context. Many authors are aware that strong empirical regularities are very difficult to be established in cognitive science. Such regularities are often formulated without a higher degree of precision when compared to many physical laws. Accordingly, in the recent literature of cognitive science there are few mentions of the word 'law', contrarily to the word 'mechanism', which appears frequently. According to Bechtel and Wright (2009, p. 117), psychologists rarely establish laws to explain their phenomena, except in psychophysics. In order to support this claim, bibliometric studies are sometimes cited. For example, Teigen (2002, p. 103; cf. Roeckelein 1996), based on a bibliometric research in *PsycLit*, argues that in mainstream psychology of the twentieth century few psychological 'laws' have been proposed and that the number of citations of laws has been decreasing throughout the century. He concludes that this could be the result of increasing doubts about the lawfulness of psychological processes. In this case, one would argue that psychology (cognitive science) should not be considered a nomothetic science anymore. Similarly, in a review of works in the field of memory, Roedinger (2008, p. 225) argues that for 120 years cognitive psychologists tried to discover general laws of learning and memory, but none has stood the test of time. The central lesson to be gained from thousands of experimental studies, in the author's view, is that no general laws of memory exist (Roedinger 2008, p. 227).

Contrarily to this view, one can argue that there are, though, no bibliometric studies for the concept of 'mechanism' and this makes the comparison to be extremely difficult. This argument is, nevertheless, week and can be easily dismissed. A simple search in scientific databases such as *PsycInfo*, *Philpapers*, *Scopus* and *Web of Science* and popular textbooks of cognitive science shows that the terms 'psychological mechanism' and 'cognitive mechanism' are currently used much more frequently than 'psychological law' or 'cognitive law'. But one can argue, nevertheless, that even if the concept of 'mechanism' is indeed more frequently used in cognitive science than the concept of 'law', this does not mean that cognitive scientists are giving up on laws or that explanations in cognitive science do not seek to establish psychological laws. It could be the case that cognitive scientists do not use the particular term 'law' but, in reality, this is what they ultimately are seeking. That is, even if cognitive scientists do not understand their science as seeking for laws, this does not mean that they are not doing it.

The major problem in this debate is that there is a great amount of discussion about what exactly a 'law' is. Laws can take the form of generalizations from empirical regularities between given variables or be so general as to be considered fundamental laws of nature, such as Newton's laws. The debate concerning what exactly constitutes a 'law of nature' is vast and the demarcation criteria sometimes offered in the specialized literature to distinguish between 'laws' and 'non-laws' are a matter of major controversies.

The standard conception of 'scientific laws', given by some logical empiricists of the twentieth century, holds that these laws are universal empirical generalizations (confirmed by some positive stances of empirical evidence) that can be characterized using the logical framework of first order predicate calculus (cf. Mitchell 1997, p. S469). They do not refer to any place, time, or individual and they have counterfactual force, i.e. they support counterfactual conditions (Sober 1997). Such universal laws can be formulated in terms of propositions or abstract formulas, and they cannot be false—they can never fail to be true: be in the past, present or future, i.e. they are necessarily true, and not accidentally true. These universal generalizations are considered completely certain and absolute commandments of nature, where no exception can be admitted (Ayer 1956). Thus, they provide certain (thus, reliable) expectations and predictions of occurrence of events in nature. Since these empirical regularities between given variables are universal, they are considered fundamental laws of nature (or strict laws—S-laws).

In cognitive science, contrarily to physics, there are indisputably no fundamental universal laws, i.e. no S-laws. It is clear that generalizations in psychology are not as strong as the generalizations in some fields of physics. Psychological laws have a largely reduced scope in comparison with those fundamental laws in physics. But even if they have a less general scope, can they not be still considered laws?

In an influential paper, often cited by neo-mechanists, the philosopher Robert Cummins argues forcefully against the DNTSE. He claims that explanation under this theory is a matter of subsumption of empirical phenomena under general natural laws and these laws are in turn explained by deriving them from more fundamental laws. For instance, Newton's laws of motion explain the simple pendulum law (Cummins 2000, p. 117). However, in Cummins' view, no laws in the sciences are explanatory in this sense (2000, p. 119).

For the author, scientific explanations in cognitive science are not a matter of formulating laws. He claims that what is often called psychological 'laws' are not explanatory statements: they are merely descriptive—they just show that there is a regular relationship between variables, but they do not explain that relationship. One can always ask: why the relationship holds? Thus, for Cummins, what are called laws in psychology are *explananda*, not *explanans* (2000, p. 119). This is why generalized and well confirmed regularities in cognitive science are often called 'effects' of a given phenomenon (Cummins 2000, p. 119). 'Effects' in this sense, for Cummins, are *explananda* waiting for explanations. Popular examples are the *Stroop effect* (roughly, the tendency to take more time to name the colours of words, when the words and the colours do not match: e.g. the word 'green' appears in the colour 'red', instead of 'green'), the *McGurk effect* (roughly, the tendency of the auditory perception of phonemes to be affected by the visual perception of the speech: e.g. one hears the sound 'fa' due to the perception of the mouth movement, but the actual sound is 'ba'), the *Serial position effect* (roughly, the tendency of a person to recall the first and the last items in a series best, and the middle ones worst), and the *Garcia effect* (roughly, a tendency to have aversion for a particular smell or taste that was previously associated with some negative, unpleasant, reaction, such as nausea). Similarly, in Cummins view, there is in physics the *Photo-electric effect*, which is an *explanandum* and was explained by Albert Einstein. Cummins point is clearly put as follows:

> [...] no one thinks that the McGurk effect explains the data it subsumes. No one not in the grip of the DN model would suppose that one could *explain* why someone hears a consonant like the speaking mouth appears to make by appeal to the McGurk effect. That just *is* the McGurk effect. (Cummins 2000, p. 119—highlight in the original)

According to Cummins, in the case of physics there are fundamental laws. But fundamental physics is the only fundamental science and up to the task of providing fundamental laws of nature. Motion appears to be the same everywhere, and thus fundamental principles of motion can be established. At this fundamental level, laws can be universal and merely descriptive, because since it is *fundamental*, there can be no further explanation, and one can just say how things are. But the same does not happen with the other sciences which deal with particular kinds of systems. Cognitive science cannot provide fundamental laws, because it does not deal with fundamental phenomena in nature, but rather with particular capacities of particular natural entities. Indeed, all the laws and explanations in the special sciences are meant to be applied to particular kinds of systems (their particular objects of study) and not to be general laws of nature.

Thus, to sum up, in Cummins view, particular empirical 'laws' in psychology, thus, are particular empirical regularities that specify 'effects' and these regularities are descriptive, not explanatory, i.e. they are *explananda*. Psychological *explananda* need to be explained in terms of the particular characteristics of the systems with their particular capacities that give rise to particular effects. These particular systems can be explained physically and functionally in terms of their constitutive parts and their mode of organization (Cummins 2000, p. 122).

Some authors suggest, nevertheless, that what cognitive science calls 'effects' can indeed be considered, at least in some occasions, as providing explanations, and not merely as *explananda*.[3] Weiskopf and Adams (2015, p. 25) argue that one thing can be considered a phenomenon to be explained in some context (as *explanandum*), while in another it can be used to provide explanations (as *explanans*); for instance, the *Photo-electric effect* in physics was explained by Einstein, but it can be also used to explain why the hull of a spacecraft develops a positive charge when sunlight hits it. Since the effect does not mention anything about the functioning of a spacecraft, it cannot be merely descriptive, or a matter of confirmation of the proposed regularity by some given data set. It is rather a matter of explanation. Given that the effect deals with a general phenomenon involving the release of electrons following the absorption of photons, it can be used to explain particular phenomena and the operations of many artificial devices, such as image sensors and solar cells (Weiskopf and Adams 2015, p. 25).

Similarly, psychological effects or laws can also be used to explain occurrences of psychological phenomena, as in the case of *Fechner's law*. One could argue that establishing a law, even if a simple one, between variables is, at least in some cases, already explanatory in some sense. For the law states that there is a regularity between the variables that is not arbitrary—there is a systematic relation that can be established sometimes even in a form of simple mathematical equation—and this is what explains the relationship between these variables. From these simple laws, many

[3]It is correct that such 'effects' are difficult to find outside the context of experimental manipulations in the laboratories of cognitive scientists. They are usually not psychological or behavioral regularities that one observes easily in nature. However, these 'effects' are very helpful for the task of characterizing more precisely the psychological capacities being investigated and the cognitive phenomena related to them. The 'effects' produced experimentally in the artificial context of laboratories can help to understand malfunctions of the capacity as well as its normal functioning in real life. This is why they are useful.

predictions can also be made with some degree of accuracy, as well as manipulations in order to bring about some desired effect. Thus, one can ask: why this phenomenon happens in the way it does? The explanation can be given by means of mentioning a law describing an empirical stable regularity between variables, i.e. the phenomenon occurs in this particular way because there is a systematic connection between variables. At least in this minimal sense, a simple law can be explanatory; but it is of course possible and frequently necessary to further explain why the connection holds by means of more general regularities or by explaining how a related mechanism works.

Furthermore, there is the influential suggestion that psychological laws can be considered as *ceteris paribus laws* (C-laws) (cf. Reutlinger et al. 2015), i.e. they have many exceptions and are non-universal, but they refer to generalizations that can be used to describe empirical causal regularities, explain why these regularities occur and derive accurate predictions to some degree. This idea is often understood as the project of constructing the particular generalizations in cognitive science as a particular kind of laws, seeking at the same time to provide general conditions (namely, complete antecedent conditions) so that these laws can be considered as genuine as strict laws (Woodward 2002, p. 304). One of the central problems with this suggestion, however, is the requirement that all the antecedent conditions need to be spelled out in some way in order to make the C-law absolutely complete and non-vacuous, and thus legitimate (cf. Fodor 1991; Pietroski and Rey 1995). In this case, the antecedent will be sufficient for the generality required of a S-law. One of the major difficulties with such attempt is that to formulate all these antecedent conditions in a complete fashion is nearly impossible, since they can be extremely diverse in a great variety of cases (Woodward 2002).

It is clear, in any case, that generalizations in psychology are not as strong as the generalizations in fundamental physics. Following Leuridan (2010), Craver and Kaiser point out that all authors in this debate "agree that the traditional notion of a 'strict law,' the universally quantified material conditional with unrestricted scope and a good deal besides, has little application in biology and other special sciences" (2013, p. 127).

Accordingly, another suggestion for the consideration of scientific regularities in explanations in cognitive science can be constructed based in the works of Mitchell (1997), Woodward (2002) and Leuridan (2010). The general idea is that given the variety in terms of empirical generalizations found across the empirical sciences, it is more plausible to understand scientific practice as involving the construction of 'pragmatic laws' (P-laws). These laws are empirical generalizations that ground and inform a substantial portion of scientific activity. They are useful for explanation, prediction and manipulation. P-laws are stronger than mere accidental regularities, but do not need to be necessarily S-laws. These generalizations also describe stable causal regularities, but are contingent in different degrees, can involve complete determinism, high probability or low probability, and can be applied arguably both universally (thus, having no exception) or to smaller sets of phenomena.

A 'causal regularity' here can be understood as a statistical or deterministic pattern of causal dependence between two magnitudes (variables) and is related to the notion of recurrence—some event that happens only once cannot be considered a regularity. The degree of contingency of a causal regularity can be evaluated mainly according to two instances. Firstly, *stability*, i.e. a regularity is *stable* in proportion to the *range of conditions under which it continues to hold for a given class of events*—the less exceptions, the more stable. Secondly, *strength*, i.e. a regularity is *strong* if it is applied to a large size of the space-time region in which it holds, that is, *its range of validity*—it is a matter of the (wide or narrow) scope of the application; the greater the population (or the elements), the stronger is the regularity. Accordingly, a 'generalization' is best understood as a statement describing a regularity between variables and indicating the group of elements in a given class in which that regularity holds (or can be applied); and an 'extrapolation' is to apply that generalization to other kinds of classes, creating thereby a stronger generalization.

Thus, instead of the dichotomy between S-laws in physics and other kinds of regularities in other sciences, the difference between scientific laws refers to degrees of stability and scope concerning the empirical regularities. The more stability and generality a law has, stricter it is, and vice versa. The concept of P-laws, accordingly, appears to be more plausible and useful, since it connects the scientific activity among the sciences that attempt to provide scientific explanations by accounting for general and stable regularities in nature.

The concept of a P-law in cognitive science, therefore, calls into question the incompatibility between (i) the activity of formulating laws and (ii) the activity of explaining the functioning of mechanisms. In contrast to this view of incompatibility, one could rather plausibly argue that the two frameworks can complement each other on this central issue.

Indeed, both the mechanistic and the nomological frameworks have one particular central similarity. For Hempel and Oppenheim (1948/1965, p. 253) empirical generalizations across the sciences are based on causal relations, i.e. what they are advocating is a causal type of explanation which is adequate for all scientific fields that seek to provide scientific explanations. In their view, "individual events may conform to, and thus be explainable by means of, general laws of the causal type"; in this context a causal law asserts that "any event of a specified kind, i.e. any event having certain specified characteristics, is accompanied by another event which in turn has certain specified characteristics; for example, that in any event involving friction, heat is developed" (Hempel and Oppenheim 1948/1965, p. 253). Thus, a general law is based on a generalization of a relation of causation between phenomena. Science deals with causal explanations and causal analysis; it is from this basis that general regularities can be built (Hempel and Oppenheim 1948/1965, p. 254).

This is very similar to the claim that scientific explanations provide an account of how phenomena are situated in the causal structure of the world (cf. Craver 2007, p. 200). Often, the neo-mechanists focus on 'compositional causation'. However, in a biological complex mechanism there are causal interactions at the same level of decomposition (micro-functions/regularities), as well as causal interactions at the highest level of wholes (macro-functions/regularities). These horizontal causal interactions take the form of 'etiological causal processes'. Consequently, there are also causal regularities present all across complex neuro-cognitive biological mechanisms, such as the mechanism of visual perception in humans. These causal interactions and their regularities can be accounted in terms P-laws.

Therefore, if (1) laws of nature are understood as P-laws, that account for causal regularities in terms of different degrees of stability and strength, and if (2) the horizontal micro and macro causal regularities present in complex neuro-cognitive biological mechanisms can be described as P-laws, then a significant part of the activity of formulating scientific explanations in cognitive science can be considered as having nomological and mechanistic elements. Thus, the two views can be ultimately regarded as compatible concerning the central issue of providing scientific explanations for a significant number of phenomena in cognitive science. A substantial part of explanations in contemporary cognitive science attempt to establish P-laws, or to explain complex neuro-cognitive biological mechanisms, or to do both at the same time. There is no incompatibility with this, as long as one takes into consideration more correct formulations of 'cognitive laws' and 'neuro-cognitive mechanisms' in cognitive science.

Ultimately, then, the debate can lead to a matter of taxonomy. One can prefer to use the term 'law' for universal (wide-scope, without exceptions) scientific statements. In this way, we have perhaps just fundamental physics that is capable of providing such laws, and other kinds of laws are not worth being called laws. But one can take another direction and call laws simply P-laws, where the difference between laws is just a matter of degree in the generality and scope of the empirical regularities accounted by these laws. The second option appears more plausible, and at the same time it connects the scientific activity among the sciences that attempt to provide scientific explanations by accounting for general and stable regularities in nature. Thus, in the case of cognitive science, it remains possible to conceive its goal as providing causal explanations and generalizations in the form of P-laws, which are explanatory. Therefore, the neo-mechanists' central claim that all (or the majority, or even a substantial portion of) psychological explanations are not made in terms of laws, but in terms of mechanisms, turns out to be misleading, or at least a highly disputable matter.

Therefore, we can consider a third possibility for an account concerning scientific explanations in cognitive science. On the one side, Hempel and other logical empiricists held the strong claim that all explanations in science, including, thus, cognitive science, are provided through the construction of general laws of nature. The neo-mechanists, on the other side, claim that there are (almost) no laws in cognitive science and the aim of cognitive scientists is to provide explanations in terms of mechanisms. A third possibility is to claim that at least a substantial part of explanations in contemporary cognitive science attempt to both explain mechanisms and establish laws at the same time (this third way has been also suggested by Fodor 1990, p. 155; Weiskopf and Adams 2015, p. 28, but I elaborated the view here).

This third possibility calls into question the incompatibility between those two frameworks concerning the central aim of explanations in cognitive science. In contrast to this view of incompatibility, one could rather plausibly argue that the two frameworks can complement each other on this central issue. Indeed, both frameworks have many central similarities. Interestingly, while in many papers Bechtel criticizes vigorously the DNTSE because of the suggestion that scientific explanations in the biological sciences are provided by establishing laws of nature, in another one he endorses the third view that scientific explanations in terms of mechanisms and laws are rather compatible. The author points out that appeals to laws and appeals to mechanisms are "quite compatible", and "the mechanistic framework allows for both" (Bechtel and Wright 2009, p. 126).

Hence, the two frameworks can be regarded ultimately as compatible concerning the central issue of providing causal explanations and generalizations of these empirical regularities. Therefore, a third possibility for conceiving explanations in cognitive science appears more consistent and more plausible.

(B) *Autonomy of Psychological Explanations*

A second issue concerning psychological explanations refers to whether functional explanations—i.e. explanations of cognitive capacities in terms of functions, sub-functions and their relations—need to be considered as a kind of mechanistic explanation or whether functional explanations in cognitive science can be considered valid explanations on their own.

Influential proponents of the mechanistic framework attempt to transform traditional explanation in cognitive science into mechanistic explanations (cf. Craver 2007, p. 111). As Piccinini and Craver state, "functional analysis is [...] a kind of mechanistic explanation; a fortiori, functional analysis is not autonomous from mechanistic explanation, and psychological explanation is not autonomous from neuroscientific explanation" (2011, p. 285). Or as Boone and Piccinini put it: "The old view of psychology as autonomous from neuroscience [...] has been effectively supplanted by a new framework where multilevel integration rules the day" (2015, Sect. 6). In Craver's view, functional analysis provides an appropriate starting place for constructing an adequate account of mechanistic explanation; however, since it abstracts away from the actual details of the real structure of a mechanism, it is inappropriate as a mechanistic explanation (2007, p. 122). In order to provide appropriate mechanistic explanations, functional analyses need to be elaborated, transforming thus possible models and sketches of mechanisms into plausible models and schemas of mechanisms. Accordingly, the mechanistic framework argues that functional analysis is just a part of mechanistic explanations and need to be complemented by an account of the neural structure underlying the cognitive function in order to be considered a sound scientific explanation in cognitive science (cf. Chap. 2, Sect. 2.1.7).

Functional analysis, as developed for instance by Fodor (1968) and Cummins (1983), and present in CTHC and BDTHC, are intended to show that most cognitive capacities can be analyzed in more basic ones, and thus explained by them. The main idea is that a very complicated cognitive process can be composed by more basic cognitive processes combined. The combination of the functional components of these processes in a particular way can provide a psychological-functional explanation for the cognitive capacity being investigated. The matter being disputed is thus the following: do theories in psychology need to indicate where the capacities are localized in the brain and explain how the 'cognitive processes' are produced by neural networks? Or just the functional analysis is enough to be considered a good psychological explanation?

Let us consider five examples in which psychological explanations of a given behavior might be required. The first example is taken from the work of Mele (2013, p. 783). The author asks us to think about a person named Ann, who thought long and hard—and consciously—about whether to propose marriage to Andy. This person had already experienced two painful divorces and now is very cautious about marriage. On the other hand, she also worries about missing a great opportunity. Ann reflected about the issue for many days and felt very unsettle about it. Sometimes she was busy with work and could not think so deep about the matter. After some days, she could finally devote some serious time for deep and intense conscious thought about the issue. In the end, she decided not to propose marriage to Andy, and actually to break off the relationship. This is the kind of situation that humans frequently face in concrete daily live and demand high level complex cognitive capacities to get through.

The second example is an adaptation based on the work of Korsgaard (1996, p. 15). Imagine that a person named Johannes is living under the Nazi regime and a family of Jews, trying to hide, knocks his door. A father, mother and two small children. Automatically Johannes thinks he needs to help them and he immediately let them enter his house in order to hide somewhere. Johannes sees they are extremely afraid, weak due to bad nutrition, etc. After some minutes, though, when the situation is more or less under control, he starts to think that it is a danger to have them in his own house. Johannes is putting in danger his own safety and the safety of his beloved family. But Johannes also knows this is what his 'consciousness' tells him is the right thing to do. One hour after this happened, Nazi soldiers knock on his door asking for the missing family. Now Johannes has a decision to make. If he gives them to the Nazis, claiming that he was just going to report them, he probably saves himself and his family, but the Jewish family will probably die. If the Nazis find them in Johannes' house, he can face very painful torture and death. But regardless of the danger he brings to himself and his own family, Johannes chooses to keep hiding the Jewish family.

Consider, as well, the case of a very responsible and well-educated citizen named Mary who needs to decide between two major politicians disputing for being the next president of her country. In order to make a reasonable decision, Mary's cognitive capacity of informal reasoning or critical thinking needs to be used. This can be done through reflection upon, for example, the arguments given by the very politicians disputing the position, the arguments given by political commentators and journalists from the press, the arguments given by political scientists, friends, family, and people that Mary admires and relies on when it comes to political issues, and finally reflection upon her own values and system of beliefs. After considering carefully all the positive and negative aspects of all the arguments Mary was exposed to, she finally decides for what she thinks is the best candidate. And then she goes to the appropriate place at the appropriate time in order give her vote.

Another example is the case of David. He is a young man that feels it is his duty to enter the army in order to defend his country against external aggression and provide humanitarian help to foreign countries in case of necessity. After some years of service, David is sent to a mission in a foreign country and during the activities, which involved conflict and violence, his group is able to capture a person belonging to the 'enemy group'. David is ordered by his superior to torture the enemy in order to obtain crucial information that would benefit his beloved country in many ways—so he was told. David, however, thinks that to torture another human being, no matter what are the circumstances, is morally wrong. Now he faces a difficult challenge: he needs to decide between not following a direct order from his superior and thus not helping his country as he wanted, and torturing a human being, which he thinks is categorically wrong. After some reflection, David refuses to torture the prisoner.

Consider, finally, the case of Jack and Laura, a couple which, after spending some pleasant years together, just got married. After some months of marriage, the couple faces the hard decision of having a child or not. They consider consciously many related issues and think about it for a long time. They consider their financial situation, the stability that they have in their lives and their jobs, their plans for the future and their time available for the child. They consider safety and security in the place they live and whether there are good schools around. They also consider if they are really prepared for something like this, which can be a major challenge in many different respects. They consider if they really should bring a child to a world such as ours. They consider as well what their families and friends have to say about it, but they know they need to take an important decision like this one with complete autonomy. Many other important things are considered by the couple. After talking about the issues over and over, given also their complex feelings and emotions in play, and trying to figure out all the main points to consider on the topic, they finally decide that the positive aspects of it are greater: it makes all the certain troubles that come with it worthwhile. Thus, they decide to have a child together.

Now, what explains the behavior of Ann of not proposing marriage to Andy in the first example? What are the neural mechanisms responsible for the decision and the related behavior of Ann? What are the neural networks that processed information about her thoughts and decisions? Of course, we do not know the answers for a similar real case of this kind now with high-level of detail. But even if we knew, what would be the relevance of this information for the explanation of the behavior of Ann? Similarly, we can ask: what explains the behavior of Johannes? Is the explanation found in a given mechanism in the brain? Is there a neural network or molecular mechanism responsible for the production of the final decision and related behavior? The same kind of questions can be made in the case of Mary's behavior, or the behaviors of David, Jack and Laura. In all these cases we have cognitive processes that often require consciousness and high level of complex reasoning upon systems of beliefs, being these beliefs linked in many different ways with themselves and with the environment and possessing different values, such as moral beliefs, with different emotions and attitudes related to them.

In such cases of complex conscious informal reasoning and decision making, complete explanations in cognitive science often need to go beyond the immediate neural account because the ultimate causal factors of interest lay outside the neural domain. What is important to know is what these people were thinking when they made a decision, what were the factors motivating them, what were the emotions in play; not what molecules or neural networks were activated at the time, because this does not provide the required psychological explanation. A complete psychological explanation would have to consider not just the systems of beliefs of Ann, Johannes, Mary, David, Jack and Laura and how particular beliefs are related to each other, but also, for instance, how these beliefs and motivations to act were formed. The formation of such beliefs includes external factors, such as education, culture, family environment, work environment, etc. and internal factors such as traits of personality of the person, the temper, the previous beliefs on the matter, the capacity to reason upon such kind of information, etc.

This appears to be the reason why in trying to obtain explanations for the behavior of a person, people often read biographies, for example. Why Albert Einstein behaved in the way he did on a particular occasion? A biographical book with a profound investigation of his history of life can be very informative on this. Information about the mechanisms in his brain are irrelevant here. Of course, biographies are far away from a complete scientific psychological explanation, which is interested in establishing strong regularities between variables in order to explain particular behavior. But they can illustrate what matters when it comes to provide psychological explanations for the pattern of behavior of a person with some degree of accuracy. The important point is that it is still irrelevant for a psychological explanation in many cases to try to find correlates for all the complex psychological capacities in neural systems; there are theoretical and methodological difficulties with this attempt. It is still not clear what the relevance of neural information to explain patterns of behavior related to some particular highly complex cognitive functions is. At best, in such cases, the neural information can provide a complement to the explanation. But the contextual, environmental, social, cultural, historical and other external causal factors that play a role in shaping cognitive functions and behaviors related to them are much more crucial, as well as the internal system of beliefs and their relations with the internal motivations and emotions.

A point along these lines was already made long time ago by no one less than B. F. Skinner (1904–1990). In his view, there are many questions about human behavior that cannot be answered just by looking to the brain (Skinner 1990, p. 1206). Skinner states clearly that the variables of which human behavior is a function lie in the environment (1977, p. 1). He emphasizes the important role that environmental complex contingencies play in shaping and modeling human complex behavior. This means that the important information concerning how the biological environment, the culture, the social rules, etc. change the brain, the psychological states and the behavior will be to a large extent left out of the picture. More importantly, an excessive focus on neural activity could lead to the distorting idea that the major strategy in order to change complex conscious and rational human behavior in general is direct interventions in the brain, while improvements in our environments, in our cultures and social institutions, in our social practices and in our personal belief systems could be seeing as secondary and, thus, neglected.

Evidently, this is not intended as a claim that information about neural systems and their components are always irrelevant for psychological explanations of human complex capacities, because this is simply not true. Neural information can be highly relevant in many cases. For example, in order to explain a particular change in the behavior of a person given an accident or surgery that impaired neural activity, such as in the popular case of H.M. It is crucial to emphasize the importance of neuroscientific research for the development of theories in cognitive science.

Nevertheless, on the other hand, it is important to point out the limitations in the use of neuroscientific information for the development of psychological explanations of complex human cognitive functions. Currently, we do not have methodology to track the neural basis of each belief a person can hold and it is far from clear if that can indeed be done. Would be possible to intervene in a given neural mechanism in order to change a certain complex belief? For example, would it be possible to intervene in the neural mechanism responsible for the belief that the candidate of presidency *A* should be elected president and transform it in a belief that the president *B* should be elected, or vice versa? Or would it be possible to intervene in the neural mechanism that is responsible for the formation of the belief in God and transform this person in a self-declared atheist? These questions are complex and we do not seem to be yet in the position to give a meaningful answer to them. The fact is that the human brain is very flexible and plastic, and there is no point in trying to track each human belief to a neural mechanism in order to provide neural explanations for human behavior that includes conscious complex cognitive capacities, such as conscious complex informal reasoning and decision making.

Let us consider another scenario to illustrate the point being discussed here. Consider the case where the citizen named Mary in the example above has not decided yet, but is rather still in doubt concerning her vote in the major candidates *A* and *B* to the presidency of her country. When someone asks her in what candidate she is going to vote, she says: "I still don't know". Then, some surprising news arise in a serious newspaper with charges of corruption against the candidate *A*, and Mary is a frequent reader of this newspaper, so she becomes immediately aware of this. Upon reflection on the matter and of related issues, she takes the new information seriously and she finally decides that to vote in candidate *B* is the best option. The major reason is that there is no charge whatsoever of corruption against him. When she is asked now what candidate she is going to vote for, she answers immediately: "candidate *B*". What explains the psychological phenomenon of Mary's belief change? How can information about neural networks be informative here? Evidently, an informative explanation would have to mention the most important causal factor, which is the event of the corruption charges against the candidate *A*, appearing in a serious newspaper. Moreover, the explanation would have to mention that Mary becomes aware of this event, accepts it as reliable, accepts the charges as true and accurate, and now this content is present in one or some of her beliefs. In possession of this content, Mary can rationally justify herself when engaging in discussions about the topic with family, friends and other people, providing reasons for her related beliefs and her related behaviors.

However, the significance of the event given its influence on Mary is external to activities in her neural networks and are different from merely describing what is happening in terms of neural processes. The explanation would also have to account for how this new information could change a particular belief of Mary given her system of beliefs about the topic and perhaps her system of beliefs more generally (e.g. her highest moral values). Sometimes, humans act in particular ways because they rely on their particular moral beliefs. Some human beings are even capable of sacrificing their own lives to defend their ideals. What is the neural mechanism that clearly explains the formation of such beliefs and the acts that are related to them? What relevance would such information have in an account of formation and alteration of beliefs and belief systems? The answer is none, or very little. Even if it was possible to be sure about the neural substrate of this kind of beliefs, this information alone does not tell us about the content of these beliefs, about what they refer in reality, so that this neural activity is uninformative in the case one wants to explain a process of formation or changing of beliefs given a process of complex conscious informal reasoning. The description of the related neural activity here would include in the explanation some information that is not necessary. An explanation that does not mention it, therefore, is the simplest, most informative, and, hence, the best explanation.

This explanation can be (1) supported by empirical evidence concerning how belief formation and changing in similar cases occurs in populations of human beings, (2) it can be supported by an accurate and detailed psychological account of the functional dynamics and components of belief changing and formation based on how the informal reasoning capacity of humans work (as investigated by contemporary cognitive science), and (3) it can be used for manipulation, control and prediction: if one knows what kinds of information are more likely to change the content of the beliefs of a given population of individuals related to a particular pattern of behavior, one can easily manipulate information in order to form and change those beliefs. Moreover, if the relation between particular kinds of information, belief formation/changing, and particular patterns of behavior is regular and systematic for that population of individuals in a given time given particular conditions, then the general pattern of behavior can be predicted for that population. The important point here is that a psychological explanation of that sort—concerned with belief formation and belief change through a process of conscious complex informal reasoning as in the example above—does not need to mention neural activity and, at the same time, it conforms to the basic criteria of successful scientific explanations in cognitive science endorsed even by neo-mechanists themselves (cf. Chap. 2, Sect. 2.1.7).

One last example might also help to clarify the point in discussion. Some papers have shown a significant degree of difficulties with excessive anxiety in university students from many different countries (e.g., Evans et al. 2018; Jones et al. 2018; Pedrelli et al. 2014; Quek et al. 2019). What explains this anxiety? Is there a problem with some neural system in the head of these students? There might even be, in some particular cases. However, for the vast majority, that does not appear to be the case. The papers in the specialized literature suggest, for example: financial difficulties; academic distress; difficulties with planning the routine; not having time enough to relax and have fun; difficulties with the family environment; difficulties with making friends, because of the competitive environment; difficulties with the professor-student relationship; excessive worries about the uncertain future and about finding a job after college; doubts about the chosen field of study. Factors such as these often contribute to generate in the cognitive system of many university students distorted and dysfunctional negative beliefs about their lives and about themselves (cf. Beck and Emery 1985; Clark and Beck 2010; Hofmann 2012), which in many cases account plausibly for their high levels of anxiety at the university context. One can see, therefore, that often the interesting variables for psychological explanations are environmental, cultural, social and psychological (cf. Luria 1976).

The advocates of MTHC do not discuss such kind of examples. Instead, they discuss the physiology of memory consolidation (Bechtel 2009b), spatial memory in rodents (Craver 2007), the biological mechanism of circadian rhythm (Bechtel 2012), or networks employed in systems biology and neuroscience (Bechtel 2017) as good examples for building a theory of explanation for psychological phenomena and human behavior. How can a theory of scientific psychological explanation be built without a systematic study of genuine psychological phenomena, simple and complex? Current philosophers of biology and neuroscience complain that former philosophers of science were imposing a view on biology and neuroscience based on a theory of scientific explanation constructed with examples of explanations taken from the physical sciences. In the same way, one could argue that current philosophers of biology are imposing a view on cognitive science based on scientific explanations taken from biology and neuroscience.

Many explanations of complex human behavior in psychology need to proceed in an independent way. If the aim is to scientifically explain conscious and intelligent behavior of sound human beings, as cognitive science aims to do, often a sound scientific psychological explanation has a strong autonomy regarding neural theories. This means that many scientific explanations in psychology do not conform to the particular norms of mechanistic scientific explanations provided by the mechanistic account, since these norms require some misguided integration of cognitive science and cognitive phenomena to neuroscience and neural activity. The mechanistic account is misguided because it does not take properly into consideration the nuances of scientific explanations in cognitive science and the particularities concerning certain complex human cognitive capacities.

It is possible that in cognitive science different kinds of scientific explanations are necessary to account for the diversity of the phenomena. Ultimately, our conception of the nature of human cognitive capacities and the nature of scientific explanations in cognitive science can influence our understanding of how many kinds of explanations are needed in order to account for all the diversity of phenomena found in the field. However, certainly there are genuine and paradigmatic examples of human cognitive capacities that cannot yet be accounted by MTHC, and to this extent they present major challenges to the ambitions of the proponents of the framework. To demand from all psychological explanations that for being sound and successful they need to explain as well how the psychological functions are performed by neural systems can be extremely misleading and completely distort the purpose of many psychological explanations of complex human cognitive and behavioral capacities in the contemporary field of cognitive science.

Instead of viewing explanations in cognitive science as being all mechanistic or all non-mechanistic, an alternative view can be more plausible and more fruitful. The diverse kinds of psychological explanation in cognitive science can be understood as varying across a continuum, where for more simple psychological phenomena, such as visual perception of edges and motion in particular areas of the brain, one has complete mechanistic explanations, but, when psychological phenomena is more complex and involves conscious processes such as conscious complex informal reasoning and decision making, explanations can be purely psychological (not mechanistic) and do not need to mention information concerning neural activity. This view, if properly elaborated, might best account for the vast diversity and different degrees of complexity found in proposed explanations of human cognition in the field of cognitive science.

6.1.3 Challenge II: The Nature of Human Cognitive Capacities

Human cognitive capacities, as investigated in traditional scientific psychology, using theories such as BDTHC, and in traditional cognitive science (taken in the strict sense), using theories such as CTHC, are something evidently hard to precisely characterize, and this difficulty cannot be underestimated. It is this enormous difficulty that gives rise to different usages and applications of the concepts of 'mind' and 'cognition'. This problem with the correct characterization of the central object of study also gives rise to divergence concerning fields of inquiry. Thus, there is *philosophy of cognitive science* (cf. Clark 2014; Margolis et al. 2012) and *philosophy of psychology* (cf. Bermúdez 2005; Weiskopf and Adams 2015). Similarly, there are books on *philosophy of mind* (cf. Kim 2011) and *philosophy of cognition* (cf. Braddon-Mitchell and Jackson 2007). In the sciences, there are books on *psychology* (cf. Kalat 2011), *cognitive psychology* (cf. Anderson 2015; Eysenck and Keane 2015; Reisberg 2013), *cognitive science* (cf. Bermúdez 2014; Chipman 2017), and *cognitive neuroscience* (Ochsner and Kosslyn 2014; Ward 2015) allegedly dealing directly with human 'mind' or 'cognition'. The differences between these fields are far from clear in the literature, as well as their more precise characterizations, scopes and limits.

Moreover, the characterizations of the science and of its object of study in the relevant books do not obey any particular or general pattern and appear as different as it can be in different works. Let us consider a very tiny sample to illustrate. Weiskopf and Adams state that "psychology" is the "science of the mind" (2015, p. ix). Clark suggests that cognitive science is the "scientific study of the embodied, environmentally embedded mind" (2014, p. vii). Bermúdez claims that cognitive science is the science of human mind (2014, p. xxvii), or science of mind (2014, p. 3). Harnish claims that cognitive science is the scientific study of cognition (2002, p. xv). In Thagard's view, cognitive science is the science of mind and intelligence

(2005, p. ix; 2018, intro.). Gardner claims that cognitive science is the science of human knowledge (1985, p. 6). Anderson claims that "cognitive psychology is the science of how the mind is organized to produce intelligent thought and how the mind is realized in the brain" (2015, p. 1). In the view of Sternberg and Sternberg, "cognitive psychology is the study of how people perceive, learn, remember, and think about information" (2012, p. 3). Eysenck and Keane claim that cognitive psychology is the science of human cognition, behavior and brain activity (2015, p. xv, 1). Goldstein states that cognitive psychology is the science of the mind (2015, p. 4). Occasionally, we encounter also authors using new terms such as 'mind-brain', or 'mindware'. Are all these authors talking about the same thing with different words? Are they talking about completely different things? Are they talking about different aspects of the same thing? Are they talking about different things but that have many aspects in common? Do we have any well-accepted criteria to decide? This confusion is a reflection of the theoretical difficulties that one finds almost everywhere concerning the topic of human cognition.

Does MTHC provide any rigorous criteria to cast some light on this confusion and provide unification concerning what (human) mind or cognition is? It is very hard to answer this question with a 'yes'. Bechtel (2008) uses the concepts of 'mental mechanism' and 'mind-brain' and is more sympathetic towards cognitive neuroscience. But why not consider cognitive neuroscience a subfield of cognitive science, for instance? Craver (2007), on the other hand, considers cognitive psychology as a part of neuroscience. But why exactly would cognitive psychology be a sub-field in the general field of neuroscience? Boone and Piccinini (2015) support cognitive neuroscience. But Thagard (2018) prefers cognitive science. How can these very different views of the neo-mechanists provide any integration for a field that is already so theoretically confuse and fragmented?

Another related difficulty is that cognitive phenomena are extremely diverse. In the literature one frequently finds terms such as 'perception', 'attention', 'memory', 'language', 'thought', 'reasoning', 'consciousness'. But sometimes also 'emotion', 'feeling', 'sensation', 'motivation'. As well as 'belief', 'intention', 'desire', 'hope', 'fear', 'frustration', 'anxiety', 'guilt', 'pride', 'shame', 'pain', 'pleasure', 'imagery', 'action', 'learning'. And even 'dreams', 'unconsciousness', 'love', 'hate', 'happiness'. It is virtually impossible to be exhaustive. Furthermore, these phenomena are related to each other in a variety of ways, as well as with things that we do not usually consider to be mental or cognitive, such as 'brain activity', 'body', 'behavior', 'computation', 'environment', 'society', 'culture', 'technology' and so on. What we do with such astonishing diversity? Of course, we try to sort them in comprehensive inter-related categories. Then, once more, there is overwhelming disagreement. Sometimes, though, three general categories are mentioned: (1) 'intellectual phenomena', which includes beliefs, intentions, desires, hopes, plans; (2) 'sensory phenomena', such as the sensation of smelling a particular odor, or the sensation of hearing a particular song; (3) phenomena related to feelings, such as the feeling of pleasure or displeasure, the feeling of fear, hate, and more complex emotions such as frustration, sadness, anger.

Unsurprisingly, such categorizations generate a vast amount of controversy, challenging any sort of unification in the field. For example, some authors would use the term 'cognitive' to describe the first category, 'intellectual phenomena', using it thus in a more restrict sense than 'mental'. These terminological and conceptual controversies on the most central concepts remain unsolved and generally unaddressed in the field of cognitive science and related scientific and philosophical areas. Furthermore, some authors would also describe this category as related to what is normally called 'higher-level cognitive phenomena', but what this higher-level means is difficult to clarify. Does it mean that perception and memory are in a lower level than planning and making decisions because they involve less rationality of some sort and this rationality is being used as a single criterion to classify psychological phenomena in the categories of 'higher' and 'lower level'? Or is it a matter of complexity, i.e. that perception or attention are less complex psychological phenomena than deductive reasoning and problem solving? Or is it something else? Finally, how to relate complex phenomena such as happiness, or love to these categories and how to analyse them in a meaningful way? More important: what is the concrete unification that MTHC provides concerning such issues of human cognition in cognitive science and related scientific and philosophical fields? So far, not much.

Instead of attempting to provide some unification in this sense, many neo-mechanists, when discussing about psychological phenomena, use examples such as circadian rhythms, or the action potential process in neurons, or something similar, as mentioned above. Occasionally, examples concerning more genuine cognitive capacities are used, such as visual perception and episodic long-term memory; however, most of the discussion focus almost entirely in neural (molecular and cellular) information provided by neuroanatomy and neurophysiology, or in behavioral experiments made with nonhuman animals. There are no major treatments by the neo-mechanists of genuine well-recognized successful psychological explanations of genuine psychological phenomena related, for instance, to traditional or contemporary developmental psychology, or social psychology, or educational psychology that appear in major journals of contemporary psychology. Deep and systematic discussions about human capacities of being conscious of external events of the environment and internal psychological states are not to be found in their works. Similarly, it is hard to find detailed discussions about the human capacity of self-consciousness and the unity of consciousness that appears to pose an obstacle to any sort of mechanistic functional and structural decomposition. The same happens in the case of systematic discussions about the normative component of human moral beliefs and its compatibility with mechanistic explanations of human cognition. These discussions are not found in the major books and papers of the major proponents of MTHC. Without such treatment, which should include a detailed examination of a reasonable number of examples of genuine psychological phenomena taken from different genuinely representative fields of traditional and contemporary psychology, MTHC appears extremely limited in scope and evidently not able to provide the unification that its advocates claim it provides. Sometimes it feels like as if the neo-mechanists were throwing a party to commemorate the unification of cognitive science, finally, but they forgot to invite the psychologists.

Let us examine, nevertheless, the more particular contributions of the mechanistic theory concerning the nature of human cognitive phenomena. At least four major controversial issues appear frequently in the contemporary literature of the major scientific and philosophical branches interested in human cognition: (A) the multiple realizability of cognitive phenomena; (B) the identification of cognitive phenomena with physical (neural) phenomena; (C) the use of cognitive representations to characterize cognitive phenomena; and (D) the use of the notions of computation and information processing to characterize cognitive phenomena. I will investigate where the advocates of MTHC stand concerning these issues and whether they are able to provide any sort of integration/unification by presenting plausible and sound views on these highly complex and difficult topics.

(A) *Multiple Realizability*

One of the main arguments that has been proposed in order to argue for the theoretical autonomy of psychological explanations in the light of neuroscientific explanations of neural activity—with important implications for a particular understanding of the neuro-cognitive relationship—is the argument from *multiple realizability* (*MR*).

This argument was initially formulated in the decade of 1960s by Hilary Putnam (1967/1975) and later developed by, among others, Fodor (1974). There is an extremely vast and rich specialized literature written about this argument and it is still vigorously debated in present days. It has been very significant for a variety of discussions in the fields of philosophy of mind, philosophy of cognitive science, cognitive science, cognitive neuroscience and beyond. Thus, my analysis does not intend to be exhaustive, evidently. The purpose is to provide the central idea of the argument and discuss the implications of it for the aims of the advocates of MTHC.[4]

The argument affirms that a given cognitive kind (states, properties, capacities, processes, functions) can be realized by many different physical kinds. For instance, the cognitive state of 'being in pain', or 'having a particular kind of belief', can be realized by many different physical structures, e.g. different types of neural networks. Therefore, according to the argument, types of cognitive states cannot be identified with types of physical states. Thus, the cognitive state of 'being in pain', for example, is not identical to, let us say, a specific type of connections between specific types of neurons in the human brain, because it is very likely that this type of cognitive state can be realized in brains of organisms that do not have this specific type of neural connection and yet the cognitive state of being in pain would remain the same. In order to make the neuro-cognitive type identity theory collapse, it is enough to find just one cognitive state that remains the same in different species, but whose physio-chemical 'realizer' is different in these species. In Putnam's view, this seems extremely likely since states like pain seem to be realized in different ways in different species (e.g. the brain of a mollusk, in his view, can likely realize the same pain the human brain does, despite being very different—cf. Putnam 1967/1975, p. 436).[5] The same would

[4]For a more comprehensive discussion cf. Bickle (2020).

[5]This assumption has been highly questioned, though. It is not clear whether there is identity between human pains and other animals' pains. Equally, it is not clear whether there is identity between many cognitive functions in humans and other cognitive functions in other animals.

be true for many others, or perhaps for all, cognitive states, which could be realized by different 'types of brains' or by a machine with the same capacity of realization that the brain has. Therefore, cognitive states cannot be identical to neural states and cannot be reduced to them in this way. After all, there is no need to assume that only brain states can realize mental states: why should the brain be the only physical basis by means of which cognitive states can be realized? Computers can realize many processes of calculation that are accomplished by humans through reasoning. As Putnam states: "We could be made of Swiss cheese and it wouldn't matter" (1975, p. 291).

The *MR* argument also conforms well to the neuroscientific discoveries that specific systems in the brain are often flexible and independent. Systems can be changed or replaced, being only necessary that they are compatible with the general architecture of the whole greater system of which they are a part, and that they still perform the same function, i.e. are responsible for realizing the same cognitive function. For example, consider a brain as a whole system and regions of the brain containing connections of neural networks as subsystems of the whole system. Consider also that these regions (subsystems) are responsible for performing some different cognitive functions. Then, the idea is that in case some region in this brain is damaged, it could be replaced by another one, or even an artificial one, when technology permits, as long as the new subsystem remains functional as before (having the same causal role, producing the same effect) and compatible with the entire structure of the brain. And the substitution could occur also if the other part was in some respects different from the original subsystem, i.e. the original neurons and their respective connections, as long as the function is performed in the same way. Accordingly, a human cognitive function related to a specific behavior can be implemented or realized in many different forms in the human brain.

The argument was also especially important for the classic program of research in cognitive science, where many authors advanced the so called 'computational metaphor'. The popular metaphor treats human cognition as analogous to a computer software and the human brain as the hardware. At the same time, it gives cognitive psychological explanations autonomy towards neuroscience, because, following the same idea present in the *MR* argument, one can argue that the same software can be realized by many different types of hardware. Cognitive phenomena, accordingly, could be realized in a brain, in a machine, or in some other similar structure. However, the structure was not considered to be important. Historically, neuroscience was part of the six disciplines that constituted traditional cognitive science, but its role was to focus on the structure and the implementation, not so much on cognition. This division created two general levels of inquiry: (1) the higher cognitive/functional/representational/computational level; and (2) the lower neural/structural/implementational level. In this traditional view of cognitive science, all that matters is the rules and functions related to the software. Thus, just cognitive functions themselves and their relations are interesting to figure in psychological explanations.

Accordingly, the *MR* argument has been frequently used by many authors in science and philosophy in two important ways. Firstly, in order to provide support for the classic functionalist view of cognition in philosophy of mind. As Fodor states: "The conventional wisdom in the philosophy of mind is that psychological states are functional" (1997, p. 149). Secondly, the argument is used to support some kind of irreducibility thesis of cognitive phenomena and theories in cognitive science. For example, Fodor claims that "the laws and theories that figure in psychological explanations are autonomous" (1997, p. 149), and that "the assumption that the subject-matter of psychology is part of the subject-matter of physics is taken to imply that psychological theories must reduce to physical theories, and it is this latter principle that makes the trouble" (1974, pp. 97–98). And the reason why neuro-cognitive reduction is troublesome is *MR* of cognitive states.

As Fodor clearly affirms: "I am strongly inclined to think that psychological states are multiply realized and that this fact refutes psychophysical reductionism once and for all" (1997, p. 149; cf. 1974). The same strategy is used by other authors. For example, Van Gulick (1992, p. 157) affirms that: "Despite initial optimism, we have not been able to reduce mental states to behavioral dispositions, nor to type-identify mental properties with properties specified in the vocabulary of neuroscience." And this is how Block starts one of his papers: "For nearly thirty years, there has been a consensus (at least in English speaking countries) that [type-type mind-brain identity] reductionism is a mistake and that there are autonomous special sciences. This consensus has been based on an argument from multiple realizability" (1997, p. 107). Many other authors have used the argument with the same intentions, and it remains one of the most influential on the debates about reductionism and non-reductionism of neuro-cognitive phenomena and theories in cognitive science and the related philosophical arena.

Towards the end of the twentieth century, however, influenced also by the new methodologies of brain research, many scientists and philosophers started to question the autonomy of psychology strongly based on the *MR* argument. One of the central critiques was, roughly, that computational artificial cognitive systems do not work as natural cognitive systems and information about neural activity is indispensable for understanding the natural realization of cognitive functions. Neuroscience and cognitive science should work in an integrative way and not in complete autonomy from each other. The *MR* argument was then highly questioned and transformed in a central target.

MTHC was developed following this trend (cf. Chap. 3, Sect. 3.1.1), and some of its advocates have unsurprisingly made a strong case against *MR* of cognitive phenomena. In Bechtel's view, for instance, evidence from neuroscience (comparative studies with brain-damaged animals, PET and fMRI) shows that there are many relevant similarities in brain areas of humans and across different species of non-human animals; he argues that if cognitive states were described in terms as fine-grained as those used to describe neural states (e.g. if we hypothesize that a specific neural mechanism produces a specific state of pain, rather than pain in general), multiple realizability is less plausible (Bechtel 2008, p. 139; cf. 2012, p. 44; cf. Bechtel and Mundale 1999). A single mechanism responsible for producing a specific cognitive state (e.g. a human pain state or a mollusk pain state), with all its particularities, can be localized; hence, local reductions through (heuristic) identity relations (involving more equivalently grained states) are more plausible. In other words, as Bechtel points out, if one uses the same standards of typing for cognitive and neural phenomena, in terms of fine and coarse grain, then "types might range across species and enable scientists to claim that the same type of mechanism in different species produces the same type of [cognitive] phenomena" (2012, p. 45). Accordingly, Bechtel concludes: "Type identity claims are core to the practice of mechanistic explanation in biology [and cognitive science] and are not jeopardized by the philosophical claims of multiple realization" (Bechtel 2012, p. 62).

Furthermore, Bechtel and Mundale write that the appeal to psychological functions is essential to brain mapping practices and functional localization, which are frequently carried out comparatively across species, but if *MR* of psychological functions occurs "brain taxonomy would have to be carried out both independently of psychological function, and without comparative evaluation across species" (1999, p. 177). Since this does not happen, as they argue, *MR* must be false. Moreover, in their view, data concerning functional localization of psychological functions provided by techniques of neuroimaging such as PET and fMRI provides evidence that *MR* does not happen. Since the localizations of psychological functions are the same in the brains of different individuals (e.g. the brain area for visual perception), there can be no multiple realization of these functions, they argue.

In arguing against *MR* of cognitive kinds Bechtel joins a group of authors that share the same intention, e.g. Kim (1992), Shapiro (2000), Polger (2009), Bickle (2003, 2006, 2012), Polger and Shapiro (2016). Bickle, for instance, claims that: "Multiple realizability [...] no longer creates serious problems [for reductionism] because evidence is accumulating for significant identity rather than diversity among the molecular mechanisms of cognitive functions shared across species" (2012, p. 105). He claims further that the same cognitive function of memory consolidation, for example, is performed by identical mechanisms in difference species, such as fruit flies, sea slugs, and mammals. Besides, he states that "more discoveries of shared molecular mechanisms for other cognitive functions across different biological genera will be forthcoming" (Bickle 2012, pp. 105–106). Thus, he concludes that "when one delves down into the cells and their molecular components that comprise [...] distinct [neural] circuits and systems [...], one finds identity among the causal mechanisms" (Bickle 2012, p. 106).

However, there are many authors defending *MR* of cognitive phenomena in the recent literature, e.g. Aizawa (2007, 2013), Aizawa and Gillett (2009, 2011), Weiskopf (2011), Figdor (2010). Aizawa, for instance, claims, contrarily to Bickle's view, that "there are substantive reasons to think that memory consolidation is multiply realized and multiply realizable" (2007, p. 65). In his paper, Aizawa shows that the proteins constituting the biochemical pathway related to memory consolidation "consist of distinct sequences of amino acids in mammals, *Aplysia*, and *Drosophila*" (2007, p. 67). Consequently, Bickle is simply mistaken in his claim that the biochemical molecular mechanisms are identical in all these species. Rather, the empirical evidence points out to the multiple realization of memory consolidation. At the same time, this argument of Aizawa provides answers to the concerns of Bechtel (2008) and Bechtel and Mundale (1999) related to the issues of fine and coarse grain of cognitive and neural phenomena. Aizawa shows that a fine-grained cognitive function can be related to fine grained neural phenomena and there is still multiple realization. As a result, it is still plausible to argue that *MR* of cognitive phenomena can occur for different grains of functional specification (e.g. coarse-grain types of cognitive phenomena, or fine-grain memory consolidation, in particular) among different species, different individuals of the same species, or even the same individual at different moments in the lifespan.

Following a similar argumentative line, Aizawa and Gillett claim that there is "overwhelming scientific evidence for what we call the massive multiple realization (MMR) hypothesis about psychological properties", i.e. many "human psychological properties are multiply realized at many neurobiological levels" (2009, p. 540). This is so, because at each level of structure and organization in neuroscience—from small individual proteins to large communities of interacting organisms—there is enormous individual variations. Aizawa and Gillett (2009, p. 540) point out that organisms, in fact, vary not just in their genetic makeup, but also in their history of interactions with the environment; thus, there are no organisms that are exactly alike.

Accordingly, "for many human psychological properties, the instances of these properties are realized by different lower level properties at many of the levels studied in neuroscience" (Aizawa and Gillett 2009, p. 540). This evidence of individual variation, from neuroscientific levels of proteins to whole brains of organisms, shows that massive multiple realization occurs and that common objections to *MR* in cognitive science are mistaken.

Aizawa (2009) provides as well a direct reply to the arguments against *MR* offered by Bechtel and Mundale (1999). He shows in his paper a series of flaws in their line of argumentation and problems with their interpretation of procedures and evidence in neuroscience. More particularly, Aizawa shows that Bechtel and Mundale "misrepresent the nature of anatomical brain mapping", "misinterpret the bearing of studies of functional localization on the issues of the multiple realization of psychological properties", and that the methodological implication of *MR* of psychological functions "are not what [they] presuppose", i.e. that *MR* of psychological functions leads to independent research on taxonomies of the brain (2009, p. 494). In Aizawa's view, it is implausible that brain taxonomy makes essential use of psychological function. This is valid for brain mapping in general and for neuroanatomical brain mapping (2009, p. 497). In fact, as he shows, it is simply not the case that "knowledge of function, any kind of function, is essential to brain mapping" (Aizawa 2009, p. 500).

Furthermore, Bechtel and Mundale conflate two very different projects in neuroscience, namely brain mapping and functional localization. The investigation of functions is essential just for the later, but not for the former. Yet, concerning psychological functional localization in the brain, the authors also commit a mistake: they conflate 'unique localization' with 'unique realization'. These are different because while unique localization means, roughly, that a particular cognitive function occurs always in the same region of the brain, unique realization means, roughly, that a particular cognitive function is *constructed* always in the same way (cell by cell, molecule by molecule, connection by connection) in that region. Psychological functions can be thus uniquely localized and at the same time multiply realized, contrarily to what the authors believe (Aizawa 2009, p. 502).

Techniques of neuroimaging such as PET and fMRI apply procedures in order to normalize and average individual brain collected data onto a single standard human brain model: in this way "an image of a subject's brain is stretched or squeezed so as to fit on an image of the standard brain" (Aizawa 2009, p. 503). These procedures are applied in order to discover commonalities and mask differences among individual subjects at the outset. In this case, these techniques eliminate the very possibility of finding *MR* of psychological functions, being thus a mistake to see them as evidence for the inexistence of *MR*, as Bechtel and Mundale do (Aizawa 2009, p. 503). Besides, the popular BOLD fMRI measures, as is well known, local changes in blood oxygenation in the brain, i.e. the commonalities it shows related to different brain regions is in terms of sameness in changes of blood oxygenation. But this is "far from showing that these brain regions constitute a unique type of realization for a given psychological function" (Aizawa 2009, p. 504). In other words, commonalities in neural metabolic activities do not show commonalities in realization of cognitive functions. The case of PET is similar: same levels of positron emissions in particular averaged regions of a standard brain do not mean same realization of psychological function in different real subjects.

Finally, Aizawa shows that psychological functions might be multiple realized and this does not prevent the mapping and categorizing of brain functions, or comparative studies across species, taking into consideration psychological functions. In other words, *MR* of psychological functions does not preclude methodological and intertheoretical exchange between cognitive science and neuroscience. Cognitive scientists and neuroscientists can still map and categorize brain regions according to psychological functions, as well as make comparative studies, without problems even with multiple realizability of psychological functions. Bechtel and Mundale indeed provide no reason for thinking otherwise. Neuroanatomical similarities and differences can be discovered in any case. Multiple realizability would indeed not make impossible comparisons of the *cerebellum*, of the *corpus callosum*, of brain weights, and of neuronal cells, for example, in different species, since *MR* of psychological functions occurs in physical biological substrates (Aizawa 2009, p. 506). The discussion of the particular psychological function of memory consolidation in different species carried above provides a fine example of this.

It is also very important to point out that while some of the advocates of MTHC attack the *MR* argument, other influential advocates appear to endorse at least some versions of it. Piccinini and Craver claim that indeed "there is a kind of multiple realizability", since "there is evidence that the same psychological capacity is fulfilled at different times by entirely different neural structures, or different configurations of neural structures, even within the same organism" (2011, p. 285).[6] In their paper, the authors advance, though, a very unusual interpretation of multiple realizability. In

[6] Yet, in another paper, Boone and Piccinini (2015, Sect. 5) appear to see standard multiple realizability in a negative light since it gives primacy to 'functions' in cognitive science explanations, and

their view, *MR* occurs because the same cognitive function can have different functional decompositions, and they further claim that it is common to think, contrarily, that cognitive functions have just a single functional decomposition. There are at least three problems with this view. Firstly, the authors do not provide detailed argument to support they claim that it is common to think that cognitive functions have just one functional decomposition. Secondly, the authors are assuming that each sub-function of a given cognitive function in a cognitive functional decomposition is performed by a neural function, and in this way when the neural function changes, the functional decomposition changes. But one does not need to assume that. One can argue that the functional decomposition provides cognitive sub-functions that remain the same while the neural structures that realize them change, as in the case of memory consolidation discussed above. Thirdly, even if the sub-functions of the cognitive function under analysis change, that is irrelevant, because what matters is that the cognitive function under analysis is being realized by different neural structures—this is all that is required for a standard view of *MR*. And the authors clearly accept the main thesis of *MR*, i.e. that the same cognitive function is realized by different neural structures.

Still, in another paper, Piccinini shows with more clarity how he sees *MR* in cognitive science. In his view, it is important to debate and clarify the argument since it "contributes to the philosophy of science and metaphysics in its own right", "sheds light on one of the central issues in the philosophy of the special sciences", and "helps clarify one of the central issues of the mind-body problem" (Piccinini and Maley 2014, p. 126). For the authors, *MR* "is worth taking seriously" (Piccinini and Maley 2014, p. 148) and it can be understood "in terms of different mechanisms for the same capacity", what favors an "integrationist perspective" (Piccinini and Maley 2014, p. 126). That is, the classic view on *MR* can be reformulated in a way that "comports well with mechanistic explanation" (Piccinini and Maley 2014, p. 148). Thus, in this paper, Piccinini is flatly endorsing a kind of multiple realizability of cognitive phenomena. Therefore, on the issue of *MR* there is no unification even between the very proponents of MTHC themselves. On the contrary, there is substantial disagreement on this central topic.

In sum, the *MR* argument has been extensively discussed in the literature of philosophy of mind and philosophy of cognitive science (and beyond) since the 1960s. It is a very significant argument for philosophers and scientists interested in human cognition, particularly, and cognition in general (i.e. including non-human animals and artificial cognitive systems)—this partially explains why much has been said about it. Indeed, many authors offered relevant arguments not only in favor of this argument, but also against it. Consequently, the debate concerning the argument is complex; and it remains open, ongoing, and very alive. Given this, on the one hand, it cannot be used to argue for the autonomy of cognitive science and human cognitive phenomena, as it was frequently employed in the past, without facing strong

---

this, in their view, is not adequate, because both structure and function are equally important. Nevertheless, no particular detailed argument is advanced against the multiple realizability of cognitive capacities in this paper.

opposition. However, on the other hand, it cannot be dismissed or endorsed easily in an attempt to advocate an integration of cognitive science and neuroscience in the form that influential advocates of MTHC aim to do.

Furthermore, the clear division among some of the most influential neo-mechanists concerning central issues such as this one related to *MR* has two important implications. Firstly, it makes difficult the assessment of the mechanistic theory, since it can have very different formulations on central topics, with consequences for other central issues related to the theory. Secondly, it makes very difficult for the neo-mechanists to claim that they are able to integrate or unify meaningfully theories in the field of cognitive science.

(B) *Neuro-cognitive Identity Theory and Physicalist Ontological Monism*

The ontological monist physicalism advocated by influential neo-mechanists (cf. Chap. 3, Sect. 3.1.3) can be challenged as well. Physicalism is still highly controversial and remains being questioned by all other non-physicalist positions (cf. Koons and Bealer 2010; Stoljar 2010, 2015). Thus, contrarily to what some neo-mechanists might think (cf. Piccinini and Craver 2011, p. 284), ontological monist physicalism is everything but "uncontroversial". There is a major controversy about what exactly physicalism amounts to. After all, how one should understand the meaning of 'physical', or the idea that 'everything in the world is physical or has a strong relation to physical things'? Is this relation 'realization', 'supervenience', 'identity', 'causation', 'grounding', or something else? The nature of a physical object or property can be determined conceptually a priori? Or should we rely on the empirical (physical) sciences to tell us? If we should rely on the physical sciences, should we rely on the current incomplete physical sciences, or some future ideally complete physical science? All these matters remain controversial, and the major advocates of the mechanistic theory do not touch any of them systematically.

Particularly, a kind of physicalism appears to be sometimes assumed in obscure ways by influential neo-mechanists. Craver, for instance, sometimes appears to endorse implicitly some sort of radical ontological reductionist physicalism concerning genuine cognitive capacities and concepts. He claims frequently that concepts such as cognitive computation, representation, and other important cognitive concepts are merely place-holders for the causal story that neuroscience will supposedly provide (cf. Chap. 2, Sect. 2.1.7). But this is a position very difficult to defend. In the attempt to explain cognitive phenomena, the *explananda* in neuroscientific fields interested in cognition, as well as in cognitive neuroscience, are formulated in terms of concepts taken from traditional scientific psychology, i.e. for example, perception, memory, attention, language, reasoning and consciousness (cf. Ward 2015). And this is different from just neural or classic computation over formal or neural representations. There is no proposal of using any other purely neuroscientific concepts that could substitute the classic psychological ones and reduce the necessity of using them in scientific explanations. Any program of neuroscience that has strong attachments with this kind of reductionism at this point of our scientific and philosophical understanding of human brain and cognition runs the risk of explaining everything except human cognition itself. At this point at least, this strategy is not going to get us anywhere towards a better understanding of human cognition and especially complex human cognitive capacities. On the contrary, it is likely to mislead cognitive science, both empirically and in its theoretical activity.

Piccinini, in turn, criticizes explicitly another form of defending ontological monist physicalism, namely by using mind-brain type identity theory. This position is criticized given that it does not "adequately [capture] the main thrust of work in cognitive neuroscience, because that work is aimed at understanding complex interplay between structure and function" (Boone and Piccinini 2015, Sect. 5). The author claims that both structure and function are required to explain cognition. In discussing the same issue, Piccinini and Craver claim, nevertheless, that: "Nothing in our argument turns on there being a metaphysically fundamental divide between functional and structural properties"; "one cannot characterize functions without committing oneself with structures and vice versa" (2011, p. 287). Thus, ultimately, the relationship between cognitive functions and neural functions performed by neural structures remain obscure in the approaches of Craver and Piccinini.

In a different argumentative line, Thagard (2018, Sect. 5.3) explicitly endorses a version of mind-brain (neuro-cognitive) identity: "The increasing integration of cognitive psychology with neuroscience provides evidence for the mind-brain identity theory, according to which mental processes are neural, representational, and computational." This theory could be a form of type-type or token-token mind-brain identity theory. However, Thagard does not offer a detailed analysis of the theory and a systematic defense.

Bechtel, as well, clearly assumes a version of the mind-brain type identity theory (cf. Chap. 3, Sect. 3.1.3). And he offers a more detailed defense of his view.[7] My aim here is to analyze it and evaluate it. That is, what are Bechtel's reasons for moving from neuro-cognitive correlations, which are overwhelmingly accepted in cognitive science, to neuro-cognitive identities? The author's defense of type-type neuro-cognitive identity theory is grounded in two pillars: (1) an attack on the argument from multiple realizability (*MR*) of cognitive phenomena; (2) the claim that some empirical evidence supports identities—not just correlations—and that these identities are *heuristic*, i.e. they are merely "hypotheses that guide subsequent inquiry", rather than "conclusions of the research" (Bechtel and McCauley 1999, p. 67).

Let us consider firstly the issue of *MR*. This is how Bechtel and McCauley put the problem: "The dominant versions of functionalism [...] reject the type identity

[7]There are at least three major arguments for this position highly debated in the contemporary specialized literature. The first one is based on the loose idea of simplicity and parsimony in science, i.e. based on the attempt to argue that physicalism is not mysterious and does not ask for the belief in a second incomprehensible fundamental kind of entity or property (e.g. Smart 1959). The second argument is based on the attempt to argue that it provides the best explanation available (e.g. McLaughlin 2010). The third argument is based on the attempt to argue that it is the most plausible form of making (at least some) mental properties, processes and other kinds have causal powers in a physical world causally closed such as ours (e.g. Kim 2005, Chap. 5; Papineau 2002, Chap. 1). Each of these arguments can be challenged in different ways, but it is not my purpose to discuss them in detail here, since this is out of the scope of the present work. My concern is only with Bechtel's mechanistic defense of the neuro-cognitive type identity thesis.

of mental and physical states, since their relations are many-to-many or at least one-to-many, not one to one. This is known as the *multiple realizability* objection to identity theory" (1999, p. 67—highlights in the original). The argument starts with the authors claiming that comparative neurobiology shows that there are comparable areas in the brains of different species of organisms (e.g. humans and lemurs), despite general differences in their brains. They claim that neuroscientific research is frequently identifying brain areas and processes across a broad range of species as belonging to the same type (Bechtel and McCauley 1999, p. 68). There is however an immediate methodological difficulty with these comparisons, recognized by the authors themselves. The difficulty is that the overwhelming majority of studies in neuroscience is done in species of rodents, dogs, cats, rabbits, or monkeys; not in humans. Bechtel and McCauley (1999, p. 69) indeed recognize that some of the most important neuroscientific methods for discovering functions of different brain areas cannot be used in humans; therefore, neuroscientists often depend on indirect methods to study the human brain. Thus, comparative studies of neural activity in the case of humans are more problematic.

Moreover, these studies only inform us about neural regions that seem to be connected with, e.g., memories, visual experiences, language understanding etc.; they do not inform us about particular cognitive processes, such as memorizing how one's grandma read good night stories, or how the house where one spent the childhood was like. At any rate, after considering neurobiological comparative studies, the authors appeal to the same argument discussed above in order to argue against *MR*. The authors refer to the already mentioned paper (cf. this chapter, Sect. 6.1.3, A) where the argument is stated for the first time, i.e. in Bechtel and Mundale (1999). There is, of course, no need to repeat the analysis here, since the argument offered is exactly the same: i.e. the problem of relating coarse and fine grain of neural and cognitive processes. Moreover, no particular scientific construction of a type neuro-cognitive identity is discussed in any detail here; no concrete example is given. Thus, we just need to remember that many contemporary authors defend *MR* of fine-grained cognitive functions regarding fine-grained neural processes, such as in the paper of Aizawa (2007) discussed above. Therefore, Bechtel and McCauley attack on *MR* of cognitive phenomena in order to defend neuro-cognitive type identity relations can be dismissed, and this form of *MR* still can be regarded as presenting a problem for this view.

Now, we can consider the second argument concerning correlations and the idea of heuristic identities. This is how the authors put the problem: "The identity theory faces another objection to the effect that empirical investigations can never establish anything more than a correlation between mental events and physical events. We shall call this the *correlation* objection" (Bechtel and McCauley 1999, p. 67—highlight in the original). In the view of the authors, this objection fails because: "Scientists often propose identities during the early stages of their inquiries. These hypothetical identities are not the conclusions of scientific research but the premises" (Bechtel and McCauley 1999, p. 69). In this way, such heuristic hypotheses are employed to guide scientific activity. Since the two *relata* do not mirror each other precisely, scientists, in their view, propose heuristic hypotheses in form of identities to guide the research of each phenomenon. Thus, "psychological research" is used "to guide the discovery and elaboration of neural mechanisms" and "neural mechanisms" are used "to develop more sophisticated psychological models" (Bechtel and McCauley 1999, pp. 69–70).

A particular example concerning visual processing is offered. In their view, this example "has involved a set of related hypothetical identities that have linked neural and psychological investigation for over a hundred years in an on-going story of progressive theoretical revision at both levels of analysis" (Bechtel and McCauley 1999, p. 70). The authors state that initially there was the identification of visual processing with the cortical area of the brain V1, but, given new evidence, that was later changed to a broader cortical area. This neurobiological revision, in turn, "inspires" revisions in the "psychological account of vision" (Bechtel and McCauley 1999, p. 70). Thus, the authors claim that identities in science, and particularly, in cognitive science, are heuristic claims established at the beginning of the research process that generally guide (inspire) further scientific research.

Bechtel is basically saying that scientific identities functions in cognitive science in the same way they function in physics, chemistry or biology: "A richer appreciation of the course of scientific research over time and of the thoroughly hypothetical character of all identity claims in science argues for a heuristic conception of the identity theory" (Bechtel and McCauley 1999, p. 67). However, this is highly problematic. Classic examples of identities in other sciences are: (1) water = H_2O; (2) visible light = electromagnetic radiation within a certain portion of the electromagnetic spectrum; (3) heat = molecular kinetic energy. Such identities hold strongly and they can last even for centuries. Who has ever substantially challenged the claim that water is H_2O, or that light is electromagnetic radiation, in the actual world we live after the physical sciences said it so?

However, consider now the following identity in the field of cognitive science: pain = c-fiber activation. This identity was proposed in the literature in the second half of the last century and in present days we know it does not hold. Other similar cases of identities in cognitive science clearly do not hold: e.g. consciousness = pyramidal cell activity. Or consider examples from identities in cognitive neuroscience following Bechtel's reasoning: visual processing and regions in the occipital cortex, or episodic memory consolidation and regions in the hippocampus (cf. Chap. 3, Sects. 3.2.1 and 3.2.2). The claim that such identities in cognitive science and cognitive neuroscience hold with the same force as in the case of the identity between 'light' and 'electromagnetic radiation' in physics is simply extremely mistaken and unconvincing.

Such regular relations named 'identities' in the case of cognitive science are fragile, obscure (given the difficulties with measurements and definitions) and can be challenged in multiple ways. The two *relata* in such cases cannot be accounted independently of each other and thus be systematically related as in the cases of the physical, chemical and biological sciences. Identities in cognitive science appear to be something else, and are not worth the name. Furthermore, in the case of cognitive science one can raise concerns about identity relations based on the *MR* of cognitive functions and based on the fact that dynamical complex cognitive processes do not permit detailed identifications of the functions performed by each part of a given mechanism. Thus, the differences between establishing identities in cognitive science and establishing identities such as that of 'light' and 'electromagnetic radiation' in physics are extreme. Identities in science are relations that were supposed to hold strongly, not something that can change from one month to the other. Therefore, one can challenge the argument that what we call identities in cognitive science are really scientific identities employed in other scientific fields. Moreover, to call those connections in cognitive science 'identities' might even mislead the work, since talking about identities should carry a strong weight and this may induce researchers not to consider other options and other components when they already expect that a given restrict function performed by a given restrict mechanism should be what he is looking for in order to construct the explanation.

Another problem is that Bechtel and McCauley claim that these identities are further justified because they "integrate" the work in cognitive science and cognitive neuroscience and because they show "explanatory and predictive success" (1999, p. 71). Concerning the issue of 'integration', it is important to point out that this is not exactly what identities do. Identities are classically used in science and philosophy to reduce, ontologically, one phenomenon to another and, epistemologically, one theory to another. The classic case is the one of heat and molecular kinetic energy (cf. Chap. 4, Sect. 4.1.1). But 'integration' here appears to be used in a very loose sense, as meaning simply that the identities connect in some way research in cognitive science and neuroscience and this connection can stimulate exchange between both fields. It is difficult to disagree with this, since it is a vague statement.

Concerning the issue of explanatory success, Bechtel and McCauley could be interpreted as endorsing (as some physicalists explicitly do), at least in this particular paper, the claim that neuro-cognitive identities must be accepted because they provide the best explanation available for the (statistic) correlations (i.e. merely a probability of a regular co-presence of variables) frequently found in cognitive science between certain neural and cognitive phenomena. This would mean that $A = B$ explains the fact that whenever and only when there is A, it is very likely that there will be B. And this would be the best available explanation for the correlations (since it relies in a simpler ontology). However, this is highly problematic. For if the best available explanation is flawed, then there is indeed no good reason to accept it (Eronen 2014, p. 580). We can have one explanation that appears the best when compared with other problematic explanations, but that does not entail that the best available explanation is indeed a good explanation. And there is good reason to believe that identities in science are not established in order to explain neuro-cognitive correlations in that way.

When the mathematical physician James Maxwell (1831–1879) established the hypothesis of the scientific identity between 'light waves' and 'electromagnetic waves', he did so not because both physical phenomena are always co-present, but rather because they share a very important property: they have exactly the same 'speed of propagation' (Eronen 2014, p. 575). More empirical evidence was presented later when experimentation confirmed that both physical phenomena share other crucial properties, such as refraction, reflection and interference. One example of such scientific identities in the field of neuroscience of vision is the hypothetical identity between a kind of ganglion cell discovered in the cat retina many years ago and a kind of ganglion cell discovered more recently in mammalian retina by a different line of research (Eronen 2014, p. 575). The identity here can be established because both different lines of research found exactly the same properties for a particular population of cells; for example, the particular way they respond to light stimuli. In such cases, the evidence can be clearly produced and compared for both phenomena, and if it is found that they do share exactly all the relevant properties, a scientific identity can be established. Therefore, what the scientific identity hypothesis explains is not why there is a co-presence of two phenomena, but much more than this: it is put forward as a plausible explanation for why two phenomena that are initially considered distinct share systematically exactly the same properties. And that the different phenomena share these properties is also clearly supported by the empirical evidence produced frequently by independent lines of research.

However, in cognitive science often this is not the case. The correlations are merely a statistical co-presence. Let us take a simple case to illustrate. For instance, one holds the belief that 'Angela Merkel is the Chancellor of Germany since 2005 until the present', and there is a 'pattern of neural activity in the brain' of the person that holds this belief so that each time the person thinks on this belief, there is supposedly a particular pattern of activation in a population of neurons in the brain of the person. What are the common properties and the independent lines of research presenting empirical evidence that there is an identity here? There are none. On the contrary, the belief about Merkel has the property of being true or false, accurate or inaccurate (properties that are important for a logical and argument analysis), while the pattern of neural activity does not. Thus, it appears that the two phenomena while correlated cannot be identical, since they do not share the same properties. Therefore, no scientific identity can be put forward in this case because it will not explain anything related to common properties independently supported by empirical research. It is also far from clear how a misguided identity such as this could provide plausible scientific hypotheses of research in the relevant fields. It is much more likely that it would misguide this research and constrain it improperly. Many other similar examples can be provided in cognitive science. And it is also important to note that for a great variety of cognitive phenomena reliable systematic statistical correlations with neural activity (well supported by many different empirical studies and also successfully replicated) were not established yet; for another great number of cognitive phenomena we still have no idea about possible correlations with neural activity.

The use of such kind of neuro-cognitive identities overall in cognitive science turns out, therefore, to be highly controversial, at best. At worse, it is extremely misleading. Accordingly, there is an important outcome for MTHC concerning the particular issues of neuro-cognitive identity relations discussed here. The outcome is that some of the major advocates of MTHC are divided on this issue, as well. While Craver and Piccinini do not intend to use type-type neuro-cognitive identity relations in their frameworks, Thagard appears to accept it, and Bechtel defends it vigorously. This has the same results as the issue discussed in the section above:

it makes MTHC a theory whose central formulation significantly varies depending on the author defending it. Moreover, it poses another challenge for the ambition of neo-mechanists to unify the work in cognitive science and neuroscience, given the fact that even among the very proponents of MTHC there is substantial divergence and disunion.

(C) *Cognitive Representation*

The issue of human cognitive (or mental) representations in cognitive science is one of the most controversial. This term is one of the most central theoretical constructs in the field, considered indispensable by the majority of researchers; yet, there are some authors that forcefully deny its significance. The term has as well a long history with a rich literature in philosophy. Cognitive representations are generally and roughly taken to be structures that carry 'information' of a variety of kinds. And some argue that they have semantic properties, such as content, reference, truth-conditions and truth-value (cf. Pitt 2020). Cognitive representations are, for instance, beliefs, desires, perceptions and images. Many argue that such representations have 'intentionality', i.e. they refer to or are about certain things and they can be evaluated with respect to properties such as consistency, truth, appropriateness and accuracy (cf. Pitt 2020, Sect. 1).

Moreover, cognitive (mental) representations are related to important issues concerning consciousness, which according to some authors is the central problem in theorizing about the mind (cf. Van Gulick 2014). For instance, cognitive representations are related to the issue of 'access consciousness', which makes information available to be used for other cognitive capacities, such as reasoning (cf. Block 1995); but also, to the so called 'phenomenal consciousness', which is the state that 'feels like' something particular (cf. Nagel 1974; Jackson 1986). Many authors defend that what all cognitive phenomena have in common is that all are 'representations' constructed through concepts that are related to the world and can be combined in particular ways to generate knowledge about the world in form of perceptions or beliefs, for example. However, many other authors disagree and claim that many cognitive phenomena are not representational but states that have only a phenomenal character: they feel like something, yet it is almost impossible to externalize in form of concepts what they feel like, e.g. what it feels like to eat chocolate, or strawberry. A unified and comprehensive theory of cognitive representations need to tell us what exactly this kind of representations are, how can they have the properties we attribute to them and how they are related to other cognitive phenomena such as phenomenal consciousness.

Nevertheless, such a detailed account of cognitive representations is not found in the work of the major proponents of MTHC, much less a detailed discussion of these difficult aspects of cognitive representational and conscious phenomena. The advocates of MTHC attempt to provide an account of mental representations in the context of a physicalist program for naturalizing all aspects of human cognition. All cognitive capacities need to be explained in terms of the natural sciences (more particularly, in terms of neural activity and some sort of computation performed by it). However, neo-mechanists hold views on cognitive representations that are substantially different from each another.

Thagard, for instance, writes about "mental mechanisms that underlie mental cognition" (2006, p. 3). In his view, there are cognitive, social, neural and molecular mechanisms that are crucial for explaining cognitive phenomena: "Current cognitive science explains human thinking using a confluence of cognitive, neural, molecular, and social mechanisms. Those mechanisms most familiar are cognitive ones, which describe the mind as operating by the application of computational procedures on mental representations" (Thagard 2006, p. 5). In this account, "brain mechanisms" are those that "involve billions of neurons organized into functional areas such as the hippocampus and various parts of the cortex"; these "neural mechanisms depend on molecular mechanisms, such as the movement of ions within neurons, the flow of neurotransmitters in the synaptic connections between them, and the circulation of hormones through the brain's blood supply", but since "human thought often involves interaction with other people, we need to attend to the social mechanisms that allow one person's thinking to influence another's"; these "social mechanisms involve verbal and other kinds of communication"; these "four kinds of mechanisms" are "useful for different levels of mental explanation" (Thagard 2006, p. 6). In this account advocated by Thagard, we note four clearly distinguished levels of cognitive mechanistic explanation. However, in other accounts of mechanistic explanation in cognitive science the levels are determined just after some empirical work is done, i.e. they are relative and depend more on the phenomenon under investigation. Besides, the social level here, which is the highest level, appears to enter in the mechanistic explanation as another component of the mechanism and not as a causal external factor of influence. Thus, it makes the notions of 'constitution' and 'causation' and their distinction obscure. As one can also note, Thagard places the concept of 'cognitive representations' in the 'cognitive level', which needs to be related to the other three, molecular, neural, and social. It is also located in the third level of Thagard's mechanistic hierarchy, from the lower to the higher-level. This account of cognitive representations is not found in the work of any other influential neo-mechanist.

Bechtel indeed presents a very different idea of cognitive representations and their place in the mechanistic levels. The author's mechanistic idea of 'cognitive representation' (cf. Chap. 3, Sect. 3.1.2) refers to the case in which an area of the brain responds when a certain stimulus is presented—this area is said thus to represent the stimulus, i.e. to carry content information about the stimulus. In addition, Bechtel advocates "a control theoretic framework so as to understand the distinctive role of informational content in information-processing mechanisms" (2009a, p. 564). Thagard (2006), however, does not take this approach. His understanding of cognitive representations does not involve control theory at all.

Craver advocates yet something else. He writes about the "processing functions of brains systems" and the "representational and computational properties of brain regions" (2007, p. 9). It is not clear what this means in his work, though. He does not endorse explicitly type mind-brain identity theory (cf. this chapter, Sect. 6.1.3, B) and often appears to suggest that the concepts of mental representations, functions and computations will simply be replaced by concepts about physical processes in neural systems (cf. Chap. 2, Sect. 2.1.7).

Piccinini's view is also substantially different from the views of the other neo-mechanists. Boone and Piccinini argue that the traditional division in cognitive science and neuroscience in two levels of inquiry, 'cognitive/functional/computational vs. neural/implementational', needs to be replaced by a view where there are "*many* levels of mechanistic organization" (2015, Sect. 1—highlights in the original). They claim that "different levels are more or less cognitive depending on their specific properties" (Boone and Piccinini 2015, Sect. 1), i.e. "every level of multilevel mechanism is both *functional* and *structural*, because every level contains structures performing functions" (Boone and Piccinini 2015, Sect. 5—highlights in the original). Boone and Piccinini use also the terms 'vehicle' and 'content/semantic information' to deal with cognitive representation, and in their view "a vehicle carries *semantic information* about a source just in case it reliably correlates with the states of the source" (2015, Sect. 4—highlights in the original). The example given is the spike trains produced by neurons in the brain area V1, which allegedly reliably correlate with the presence and location of edges in the visual field. A vehicle thus represents in case it has the function of carrying information about the source, while the information carried is used by part of the system to the extent it is causally relevant to the system's performance. For the authors, representations of cognitive capacities can occur, thus, at the level of neural networks, at the level of individual neurons, and at the level of sub-neuronal structures. So far, we have four influential neo-mechanists and four substantially different accounts of cognitive mechanisms and cognitive representations.

More recently, Zednik (2018) attempts to defend an explicitly mechanistic view applied to cognitive science and cognitive phenomena as well, including cognitive representations. However, his view also differs substantially from the views of other influential neo-mechanists. The author starts by claiming that "the most influential account of explanation in cognitive science is due to David Marr (1982)" in which there are three levels of analysis: the computational, the algorithmic, and the implementational (Zednik 2018, p. 389). All these three levels are thus necessarily required for a complete explanation in cognitive science. While the author notes that Marr's account of explanation in cognitive science has been subject of intense criticism, he believes that it is, nevertheless, "a productive starting point for discussion" (Zednik 2018, p. 389). He attempts to show, thus, that "many of the ambiguities in Marr's account can be resolved, and that its scope can be extended, by considering Marr to be an early advocate of mechanistic explanation" (Zednik 2018, pp. 389–390). As he points out:

> Although Marr's account was originally designed to capture explanations in computationalist cognitive science, the questions it identifies at each level of analysis are in fact variations on the types of questions that are asked in any research program that aims to discover and describe (cognitive) mechanisms. (Zednik 2018, p. 390)

Accordingly, the author proposes to outline and defend an interpretation of Marr's account of explanations in cognitive science based on the framework of mechanistic explanations, and following Marr he claims that all three levels are necessary to provide such explanation. Moreover, Zednik explicitly recognizes that his view substantially differs from the views presented by other neo-mechanists. Cognitive representations are placed, in this view, at the algorithmic level, below the computational level. The idea is that these representations stand for patterns of information related to the environment that are used by the mechanistic system in order to perform its activities.

As the analysis of these particular accounts show, all of their views concerning cognitive representations differ in substantial ways. Thagard presents four levels of mechanistic explanations, Zednik presents three levels, while Bechtel, Craver and Piccinini do not specify a particular number of mechanistic levels, saying that it varies depending on what is being explained. In each account of levels, cognitive representations are placed in a different form and related differently with other aspects of the framework. Moreover, Thagard, Craver, Piccinini and Zednik do not endorse control systems theory in order to provide an account of cognitive representations. But for Bechtel this is crucial. Zednik is the only one that defends an account of mechanisms in cognitive science explicitly and largely based on Marr's work.

The work of Marr is, in fact, another point of major disagreement among neo-mechanists. When Boone and Piccinini are criticizing the traditional two-level view in cognitive science, the authors state that "while cognitive scientists were perhaps less explicit about the two-level picture, something similar to this view can be found in many landmark works that came out during the heyday of classical cognitive science" (2015, Sect. 2). One of the works cited by the authors here is precisely Marr (1982). Thus, the authors are claiming that Marr's view is exactly what needs to be replaced in cognitive science, while Zednik (2018) takes the work of Marr as the most fundamental basis for constructing his mechanistic proposal in cognitive science, and even claim that Marr's framework can be considered mechanistic itself. That is already quite a problem. Yet, to complicate this debate even more, Bechtel and Shagrir defend that all three levels in Marr's account need to be maintained, because they offer "distinct contributions and methodologies" (2015, p. 312). They interpret the levels of Marr's account as being levels of analysis and not levels of mechanistic compositional organization, and then they attempt to incorporate Marr's ideas into their framework of information-processing cognitive mechanisms.

This shows the high level of substantial disagreement among neo-mechanists in the most basic issues in cognitive science. Due to examples of this kind, one wonders if the neo-mechanists, when it comes to cognitive science, indeed have more in common than the mere use of the word 'mechanism'. As this analysis shows, there is a great amount of variety concerning the mechanistic theory of cognitive representations. No unitary account can be found yet in MTHC, no main common structure is found in the framework. Consequently, the same problems for the formulation of a single unified mechanistic theory (as well as for its evaluation) and for the ambition to integrate work in cognitive science are found here.

Finally, it remains extremely controversial whether any of the proposed mechanistic theories of cognitive representations can successfully account for highly complex cognitive capacities such as those discussed above (cf. this chapter, Sect. 6.1.2, B), related to conscious complex informal reasoning and decision making. Examples such as these are not discussed in detail by influential neo-mechanists when they formulate their accounts. Instead, as already mentioned, they discuss examples such as visual perception of edges in V1, neural aspects of memory consolidation, or the phenomenon of action potential in neurons such as place cells of rodents.

However, these examples are enormously different from the properties of human complex beliefs whose content can be considered non-derived, true or false, accurate or inaccurate. They are different from the capacity of human beings to form systems of beliefs and relate them according to logical rules, constructing arguments in order to support their views, which often have an influence in their behavior.[8] It is also very different from the process of thinking about different kinds of relevant information for months or years in order to take an important and complex decision (cf. this chapter, Sect. 6.1.2, B). To put it directly: the causal relations of 'representation' that hold between the activation of some place cells in rodents and the stimulus that activate it is different from the representational often conscious syntactic and semantic structures (with their internal relations and relations with external factors) present in systems of beliefs in humans. In order to make a difficult decision, a human can take into consideration information related to plans for a very distant future, in which many counterfactual scenarios are considered. A human can wonder about what happened in a very distant past, or what could have happened, even if she/he knows what actually happened. This is very different from the relationship between a place cell's action potential process in a rodent and a particular physical location of its environment. And more importantly, complex informal reasoning and complex decision making are things that humans do naturally and frequently in their daily lives.

Thus, in cognitive science one needs to deal with highly complex phenomena. Many authors still think that human beings present major differences when compared to other animals in nature. Humans have a cumulative, complex, dynamical and elaborated culture that is transmitted through generations. They also engage in understanding and writing their own history. They have natural languages with enormous complex and refined power of expression and sophisticated grammars. Humans practice and appreciate art, such as literature, painting, cinema, theater, music. They engage in purely formal or very abstract thought, when they do mathematics, logics and engage in some religious thinking. They create juridical laws for their societies and think about morality, constructing moral systems. Moreover, humans engage in politics, science and philosophy. These particularities in the complex phenomenon of human cognition appear not to be taken substantially into consideration by influential neo-mechanists interested in cognitive science.

Another important and related issue is that influential neo-mechanists do not offer a clear discussion and a clear proposal concerning the crucial problem of 'the mark

[8] A very similar point was made by Von Eckardt and Poland (2004).

of (human) cognition'. Adams and Aizawa (2010), on the contrary, based on contemporary fundamental assumptions in traditional cognitive science and philosophy of psychology/cognitive science offer some plausible criteria. These criteria may turn out to be incorrect, but at least the authors have a clear and articulated proposal in order to distinguish cognitive processes, systems and representations, from non-cognitive processes, systems and representations. This proposal is a clear theoretical contribution for the debates. The idea can be meaningfully discussed and evaluated. On the other hand, vague claims such as 'cognitive mechanisms are mechanisms that process information or that represent' must be avoided. There are many kinds of systems that process information, but cognitive systems that process information in a 'cognitive way' is just one kind of them. Equally, there are many kinds of systems that represent information, but (human) cognitive representation is just one particular kind of it. In other words: not everything that process information process it in a cognitive way; not everything that represents information represents it in a cognitive way. The neo-mechanists concerned with cognitive science need to provide a clear proposal to distinguish cognitive mechanisms from non-cognitive mechanisms (and also cognitive representations from non-cognitive representations), telling us clearly what are the particular features of these cognitive mechanisms. This account is still missing in their framework.

Therefore, one could argue that, at best, the general neuroscientific account of MTHC concerning 'neural representations' is a misapplication of a traditional and important notion used and intensively debated in the history of psychology, history of philosophy, contemporary philosophy of mind, philosophy of cognitive science and portions of classic (strict sense) cognitive science. At worse, the use of neural representations as it occurs in particular portions of neuroscience and which is adopted by some influential advocates of MTHC is no more than a convenient, but misleading and distorted way of describing in a physical, chemical and biological way what is happening—while authors claim at the same time that 'human cognition' is being somehow explained.

(D) *Cognitive Computation*

The most influential neo-mechanists in cognitive science generally defend that cognition is a kind of computation, as well. That is, cognitive representations are manipulated according to computational processes. According to some authors, computationalism is indeed the mainstream view in contemporary cognitive science. However, there are many different accounts of what computation in cognitive science is, and how exactly it should be used to characterize human cognition and be part of scientific explanations in cognitive science (cf. Piccinini 2012). What we find in cognitive science is rather a "continuum" of views on the matter, where at one end "some notions of computation are so loose that they encompass virtually everything"; for example, when computation is defined as a process between an input and an output and everything qualifies as such states (Piccinini 2012, p. 222). Moreover, to say that computation is process of information is also not helpful. One can say that virtually all organisms in one way or another gather information about their environments, but this does not tell much about what particular computational processes are being used in order to produce a cognitive capacity and the related behavior. There are many different notions of information and "the connection between information processing and computation is different depending on which notion of information is at stake" (Piccinini 2012, p. 228).

Nevertheless, authors normally recognize three general main traditions of computationalism in cognitive science: (1) classical computationalism, which emphasizes the analogy between cognitive systems and digital computers and develops computer programs to explain cognitive capacities that do not recognize a strong relevance of neurological theories and data; (2) connectionism, which provide computational explanations of cognitive capacities that are more concerned (but not entirely) with neurobiological data, i.e. real properties of real neural networks; (3) computational neuroscience, which provides neuro-computational models based on the neuroscientific data from neurophysiology and neuroanatomy, e.g. the variety of different kinds of neurons, the role of different kinds of neurotransmitters and hormones that affect the brain (Piccinini 2012, p. 225). Of course, many authors attempt to build models that combine elements from these different theories and these theories overlap in many dimensions; but it is useful to keep them separate as general accounts of computation in cognitive science.

However, it is not clear where exactly MTHC stands in the continuum that spans these traditions. Piccinini is one of the most prominent neo-mechanists concerning these matters. He defends a "mechanistic account of computation" and argues that a "computational explanation is a species of mechanistic explanation" (Piccinini 2012, p. 230). In his view, a mechanism is a "functional mechanism" (in the biological sense), since it performs a function explained by the complex organization of the sub-functions performed by the operations of its components. The computational description needs to mirror the causal structure and functions of this functional mechanism. Computation here is "the processing of vehicles according to rules that are sensitive to certain vehicle properties, and specifically to differences between different portions of the vehicles" (Piccinini 2012, p. 230). A malfunction of this functional mechanism results in a miscomputation. Moreover, the computations can be described in abstract and functional mechanisms can be considered medium-independent, i.e. the computations are independent (with some degree of variation) of the physical media that implement them:

> Thus, a given computation can be implemented in multiple physical media (e.g. mechanical, electromechanical, electronic, magnetic, etc.), provided that the media possess a sufficient number of dimensions of variation (or degrees of freedom) that can be appropriately accessed and manipulated and that the components of the mechanism are functionally organized in the appropriate way. (Piccinini 2012, p. 231)

Neural spikes (action potentials) are the vehicles used by neural processes to make computations in the generic sense, and since the relevant computational properties (e.g. spike rates) of these vehicles are medium-independent, they can be implemented by neural tissue or by some other physical medium, for instance, silicon-based circuit. It is in this sense that the brain performs computations, i.e. in the generic sense (Piccinini 2012, pp. 236–237). However, this sense is too inclusive to be explanatorily interesting. Many systems that have nothing to do with cognition can be included here given such general criteria of computation.

The author argues further that "everyone is (or should be) a connectionist or computational neuroscientist, at least in the general sense of embracing neural computation" (Piccinini 2012, p. 243). There is, however, as the author himself recognizes, a considerable difference between connectionism and computational neuroscience. Is it not even possible to indicate which one a cognitive scientist should choose? Interestingly, in another paper, Boone and Piccinini endorse explicitly computational neuroscience and claim that "connectionism is disappearing as an independent research tradition, instead merging into computational cognitive neuroscience" (2015, Sect. 6).

Piccinini also claims that "the computational study of cognition will require that we integrate different mechanistic levels into a unified, multilevel explanation of cognition" (2012, p. 243). At the same time, however, he recognizes that "much work remains to be done to characterize the specific computations on which cognition depends"; it could be that "one computational theory is right about all of cognition, or it may be that different cognitive capacities are explained by different kinds of computation"; for him, to address these questions "the only effective way is to study nervous systems at all its levels of organization and find out how they exhibit cognitive capacities" (Piccinini 2012, p. 243). However, many cognitive scientists interested in situated cognition and classic cognitive scientists (e.g. working with versions of Bayesian cognitive science) will not agree that the "only effective way" of getting these kinds of insights on human cognition is by studying the nervous system (be it at the sub-neural and neural level, or at the level of populations of neurons). Let alone psychologists of different fields, such as social psychology and developmental psychology and philosophers of mind and of cognitive science interested in human cognition.

In none of the most influential works published by influential neo-mechanists, a detailed analysis of how to apply this idea of mechanistic computation to the variety of psychological phenomena is provided. In the end, we are left with no concrete answer about how the defended mechanistic account of computation can unify the field (i.e. be applied to all the diverse psychological/cognitive phenomena). What is the kind of computation performed by functional cognitive mechanisms? What is the kind of computation performed by neural mechanisms? Is it classical sentence-like computation, is it neural networks biologically realist kind of computation? Is it algorithmic computation? Is it Turing computation? Is it analog computation? Is it quantum computation? Without knowing more precisely what cognitive computation is, it is meaningless to discuss how many computational levels are necessary for explanations of cognitive capacities. Piccinini himself recognizes that these are difficult questions to answer:

> While it is safe to say that cognition involves computation in the generic sense, and that nervous systems perform computations in the generic sense, it is much harder to establish that cognition involves a more specific kind of computation. (Piccinini 2012, p. 237)

> [...] it is very plausible that the neural processes that explain cognitive capacities are computational in the generic sense, but it is difficult to determine which specific kinds of computation – classical, digital but non-classical, analog, etc. – are involved in which cognitive capacities. Whether any particular neural computation is best regarded as a form of digital computation, analog computation, or something else is a question that cannot be settle here. (Piccinini 2012, p. 239)

Given this, one is justified to conclude that beyond the emphasis in the importance of neuroscience for research on human cognition, it is hard to see in which concrete sense the mechanistic theory provides a substantial advancement concerning the unification of our theories about cognitive computation.

Moreover, Piccinini himself recognizes that some of the most plausible arguments for the thesis that human cognition is computation comes from the classic computational theory of human cognition. One classical argument holds that human cognitive capacities exhibit productivity and systematicity (Fodor and Pylyshyn 1988), i.e. roughly, humans can produce an indefinite number of systematically related and structured sentences in natural language, and this would require the processing of language-like symbolic representations based on syntactic structures (Piccinini 2012, p. 238). Another argument states that human cognitive capacities exhibit flexibility, i.e. roughly, humans can solve a vast amount of problems (e.g. solve mathematical calculations, derive logical theorems, recognize objects, make coffee, cook pizza, play games such as chess, etc.) and learn an indefinite amount of behaviors. This must require, so goes the argument, some sort of computation, since computers present also some degree of flexibility through the process of sets of rules and instructions by a general-purpose processing mechanism designed for different tasks (Piccinini 2012, pp. 238–239). This means that the most plausible arguments available in order to show that cognition is a kind of computation are provided by a theory that the advocates of MTHC want to oppose to. At the same time, the neo-mechanists do not offer any alternative well-articulated and comprehensive view, showing how this alternative view can be applied across the astonishing variety of human cognitive phenomena. In the case of what is generally considered 'higher-level' cognitive phenomena, such as conscious problems solving, informal reasoning and decision making, the situation is even worse.

Another controversial issue is the following. Many authors argue that classic cognitive computational theory is in direct opposition to neural networks/connectionism and computational neuroscience, i.e. these are frameworks that compete to provide the best explanation of how human cognition can be considered as a kind of computation over representations. According to Garson (2019, intro.), connectionism "is a movement in cognitive science that hopes to explain intellectual abilities using artificial neural networks", which are "simplified models of the brain composed of large numbers of units" connected according to a particular pattern; this framework provides "an alternative to the classical theory of mind: the widely held view that mind is something akin to a digital computer processing a symbolic language". Many proponents of connectionism intended to provide "a new framework for understanding the nature of the mind and its relation to the brain" (Garson 2019, Sect. 4) by paying attention to the information about the properties of real neural networks in real brains. Rescorla (2020, Sect. 4) claims as well that "connectionism emerged as a prominent rival to classical computationalism". However, occasionally these frameworks are considered by neo-mechanists to be in different mechanistic levels (cf. Piccinini 2012, p. 242; Bechtel 1994), as if it was not a matter of a direct competition, but a matter of providing explanations to different levels of organization of human cognitive phenomena. Bechtel puts the point quite clearly:

> What these considerations seem to imply is that neither connectionist nor symbolic systems are likely to provide the appropriate frameworks for modeling cognitive processes. Each is based on a framework that is appropriate to a different level of theorizing, symbolic systems for the level at which humans function as conscious rule interpreters and connectionist systems for the level of neural processing. (Bechtel 1994, p. 21).

The author thus flatly endorses the idea that the framework of connectionism/neural networks is at a different mechanistic theoretical and explanatory level than the classic framework of symbolic cognitive processes.

These are two completely different ways of understanding the relationship between these frameworks and the neo-mechanists cannot give a clear and plausible answer concerning how exactly this is to be understood. All the frameworks appear to be trying to provide explanations to the same kind of phenomena, e.g. face recognition, object recognition, detection and production of grammatical structure, problem solving, etc. Thus, how can they be meaningfully talking about different levels of organization?

Perhaps, though, the frameworks should not be seeking a unified view of cognitive computation and be applied to different psychological phenomena. It is possible that it is a matter of different levels or degrees concerning different kinds of computations. Many debates and evidence appear to suggest that for some kinds of cognitive phenomena, such as face recognition, the neural networks framework is more successful; while for other kinds of cognitive phenomena, such as language production and language comprehension, classical computationalism, with its symbol manipulation, is more plausible.

However, we need to point out clearly what would be the difference in the phenomena that makes this difference in the approaches to be the case. We would need to clarify exactly how the frameworks are related in this particular sense. General conjectures now are not helpful. It could be one thing, or it could be another, or it could even be something else. What we need are consistent theoretical accounts to relate these frameworks in a significant way so that we can advance theoretically in the field. The advocates of MTHC do not offer answers here.

It is well known that connectionist neural networks and computational neuroscience cannot provide plausible explanations for the kind of "rule based processing that is thought to undergird language, reasoning, and higher forms of thought" (Garson 2019, Sect. 4). Instead of providing a systematic discussion of this problem and make a systematic comparison between connectionism, computational neuroscience models, and classical approaches in order to say who is right and how this fits MTHC, the neo-mechanists occasionally change the debate and claim these frameworks are talking about different levels of mechanistic organization. However, no clear, well-articulated and systematic attempt to relate these frameworks in a mechanistic account of levels is made. Just some general ideas and conjectures about possibilities are offered on this particular issue (cf. Bechtel 2008, Chap. 5). Furthermore, there remain many issues concerning computation and information processing in the brain (cf. Grush and Damm 2012). For example, are just neurons and neuron spikes what we need to understand neural computation, or glia cells, for instance, can also participate? Are neurons the basic unity of neural computation or also sets of neurons forming neural-networks can be considered informational processing unities? What about sub-neuronal components: is there any information processing and computation happening at this level? And what is more important in order to understand neural computation, spiking rates or spiking duration, or both? The proponents of MTHC offer no well-accepted answer for these questions. All this indicates that there is no genuine integration here, beyond appearances and rhetoric.

To sum up, MTHC provides indeed important material and ideas to discuss systematically important central issues in cognitive science, and to this extend the theory contributes to the theoretical progress in the field. However, it is virtually impossible to claim that the theory provides any substantial unification for the field, which is extremely diversified and lacks more systematic theoretical work. Concerning the particular major problems and controversies in cognitive science discussed in this chapter, basically none of the mechanistic ideas appear to solve them in a satisfactory way. If for some authors these new mechanistic ideas are groundbreaking, on the one hand, for others, they are extremely problematic, on the other. Moreover, the framework appears to be internally substantially divided, not just on general but also on particular issues, to the extent that, firstly, the advocates of the mechanistic theory applied to human cognition need to attempt a unification between themselves and, just after that, think about the attempt to provide some sort of unification for cognitive science.

6.1.4 Toward a Truly Integrated, Systematic and Genuine Science of Human Cognition

Some of the most influential twenty-first century mechanists intend to 'unify cognitive science' and to 'integrate cognitive science with neuroscience' (cf. Piccinini and Craver 2011; Bechtel and Wright 2009). In their view, this provides a 'new revolution' in cognitive science (Boone and Piccinini 2015).

Neo-mechanists generally aim to achieve this integration by imposing theoretical elements of a philosophy of biology toward cognitive science, similarly to what logical empiricists did in the past taking physics as the primary science. For a long time, psychological phenomena and explanations needed to fit the model of the physical sciences. Recently, the mechanistic theory with its strong emphasis on biological sciences has aroused and psychology once again tries to conform to the norms of a philosophy of science constructed having another science as its basis. Bechtel himself recognizes that the new theory of mechanistic explanation was constructed by different authors using a variety of particular examples of explanation taken from cell and molecular biology and neuroscience (2009a, p. 552). Traditionally, as Fodor (1968, p. xiv) has claimed, many philosophers interested in the behavioral sciences "have been primarily concerned with helping to make psychology 'scientific' by laying down guidelines for how it can conform to practices that are alleged to characterize the methodology of more advanced disciplines". Indeed, few philosophers attempt to discuss psychological theories in particular, and, as a result, "psychological metatheory has remained seriously underdeveloped" (Fodor 1968, p. xiv). Currently, the situation is similar. What we have is virtually a philosophy of biology being extended to other sciences and, in this way, being applied to cognitive phenomena and explanations.

However, as we have seen along the chapters, the ambitious rhetoric of the neo-mechanists in cognitive science collides with many obstacles and challenges. Their point of view presents a series of shortcomings. For instance, concerning the very formulation of the theory, concerning their assessment of epistemological and ontological neuro-cognitive reduction, concerning the criteria to demarcate human cognition, concerning their account of psychological explanation, concerning the issue of psychological explanatory autonomy, concerning their notion of scientific pluralist neuro-cognitive integration, and concerning the issue of cognitive/mental representations and computation. Given this, one wonders how the mechanistic theory can provide unification to the field of cognitive science if the very framework faces enormous internal problems and disunity.

In addition, the proponents of the mechanistic theory do not seem to recognize the great amount of diversity in the field of cognitive science. In this field, there are researchers belonging to a variety of different scientific fields and adopting a variety of theoretical points of view (cf. Chipman 2017; Reisberg 2013). There is work being done to understand human 'cognition' that spans from molecular neurobiology to social, cultural, developmental and education psychology. This is indeed a great amount of variation and the advocates of the mechanistic framework are not prepared for this. Their focus is rather very limited to a portion of this continuum and they end up leaving many important works being done in psychology out of the picture. Furthermore, we also need to count here philosophers of mind, philosophers of science, philosophers of psychology and philosophers of cognitive science that also aim to understand human cognition and provide different views on the matters than those views advocated by the neo-mechanists. All this legitimates skepticism concerning loose statements about unification and revolution in the field of cognitive science.

However, the issue of unity in the sciences is a very important one, and needs to be carefully considered. These matters make a difference to the extent that the positions adopted provide "guidance and even justification for hypotheses, projects and specific goals" in science (Cat 2017, Sect. 6). Concepts and ideas related to, for instance, simplicity, complexity, unity, disunity, plurality, integration and interdisciplinarity carry out often "normative value", and "provide legitimacy, even if rhetorically, in social contexts especially in situations involving sources of funding and profit. They set a standard of what carries the authority and legitimacy of what it is to be scientific" (Cat 2017, Sect. 6). They can as well influence in matters of scientific application and extension, i.e. in healthcare, economic, and science education policies. Concerning the relationship between philosophy and science, these matters can also have a direct influence on the "sort of philosophical questions to pursue and what target areas to explore", the definition of "what counts as scientific", shaping "scientistic or naturalized philosophical projects", and the definition of "what relevant science carries authority in philosophical debate" (Cat 2017, Sect. 6).

In this context of the integration of sciences in general, cognitive science needs to follow its own autonomous and independent path. It is already time for psychological explanations and its phenomena to be taken as valuable on their own. Cognitive science cannot be considered just as a field that needs to accept normative standards from sciences such as physics or biology when these standards really do not appear to fit the diversity of successful theoretical and empirical work being done in the field. Obviously, cognitive science and its scientific normative guidelines should not be considered in isolation, because there is no doubt that other sciences, e.g. neuroscience, linguistics and computer science, extremely contribute to its progress. To advocate the autonomy of some kinds of explanations in cognitive science is not to deny the importance of knowledge from other scientific areas to the field of cognitive science. Nobody can deny the influence and significance of neuroscience to cognitive science, for example, or the influence and significance of computer science and the field of artificial intelligence. But there is no compelling reason for not treating cognitive science as capable to provide its own normative guidelines considering the work done in the field itself. These guidelines can be build taking into consideration the progress and the normative guidelines from other successful sciences, but it is equally important that normative guidelines of research in the field of cognitive science be established taking into consideration primarily a significant sample of the totality of the successful work being done in this science itself.

Interdisciplinary work needs to be highly promoted in order to maintain the integration and not let the increasing extreme specialization make researchers lose sight of the totality of the scientific system of human knowledge they are but a tiny part. Moreover, at the same time interdisciplinary work in the field is encouraged so that the *internal relationships* between the different sub-fields of the science can be developed. It is also necessary to develop theoretical work in order to establish what are the scope and limits of the field, i.e. the *external relationships* between this scientific field and other scientific fields. This can be done through the development of better theories concerning general cognition and its main characteristics, i.e. through theoretical and empirical progress in cognitive science.

In order to achieve progress faster and unification in the field of cognitive science some more concrete steps can be made. Firstly, concerning unification, which is indeed important and welcome in science. This has played a very important role in the history of science and the history of our best scientific theoretical developments. We should, thus, strive to unify and integrate our scientific theories. And this unification can be constructed in very different ways (cf. Woodward 2014, Sect. 5.4). One way of constructing it is through the creation of a common descriptive and classificatory vocabulary, especially in cases where there is no previous one, in order to provide more clarity and systematicity to the science and a common language for scientific interaction. A clear example here is the comprehensive system of biological classification developed by Carl Linnaeus (1707–1778). Accordingly, researchers in cognitive science could develop a systematic and rigorous form of referring to human cognitive capacities. Some sort of taxonomy shared internationally by cognitive scientists so that they all have the same bases to talk about their issues, i.e. they share a scientific common language. This is not an easy task, of course, since interpretation of what these capacities are and their definitions will certainly vary significantly. However, it is perhaps better to try to achieve at least some minimum consensus on the most basic vocabulary, so that it can be improved with time, or discarded when some better one is presented.

Another way of constructing unification in science is genuine physical unification. In this case, common causes are used to explain phenomena previously considered to have quite different causes. For instance, Newton's unification of terrestrial and celestial theories of motion (Woodward 2014, Sect. 5.1), with his fundamental laws of motion and his theory of universal gravitation. Newton was able to demonstrate that the same cause (a single force), namely gravity was responsible for the motion of terrestrial bodies falling free near the surface of earth and the motion of planets through their orbits. Other examples are James Maxwell's unification of theories concerning electricity with theories concerning magnetism and Albert Einstein's unification of theories concerning gravity and the motion of bodies in his general theory of relativity, where he unifies Newton's theory of universal gravitation with his own special theory of relativity. Of course, to carry out projects of unification in cognitive science in the same way they are carried in physics is misleading. However, similarly to what happens in physics and in other sciences, cognitive scientists can compare theories systematically in the field and evaluate their degree of plausibility and generality. Through such a methodical comparison micro-theories and theories in a middle level can be connected to form more fundamental comprehensive systematic macro-theories which can provide more unification to the field.

Secondly, cognitive science could also profit from a clearer separation concerning its three major areas of research: (1) theoretical cognitive science; (2) experimental cognitive science; (3) applied cognitive science.[9] This is a division that is also present in physics and helps to clarify and organize more precisely the field. This separation can be interesting in cognitive science because in the field we frequently find a mixture between researchers and professionals interested in theoretical, experimental and

[9]For the sake of clarification, it is important to point out that this is not a discussion about what scientific areas belong or should belong to cognitive science, e.g. whether cognitive psychology and cognitive neuroscience belong to cognitive science or constitute separate fields. I take cognitive science in the broad sense to be the most legitimate contemporary science of human cognition, what belongs or not to it is a matter of great dispute, though (cf. Chipman 2017; Reisberg 2013). The discussion here, however, is about a major division in a scientific field of cognitive science for the sake of scientific and academic organization. It is a division concerned with the theoretical, experimental and applied dimensions of a given scientific field.

applied dimensions of the field. This mixture often creates general confusion about boundaries and interrelations between these dimensions. Of course, these dimensions are completely interrelated and exchange between them needs to be promoted, but a formal separation of them could help the progress of the science in many ways. For instance, the value of rigorous theoretical work in cognitive science can be better recognized and researchers working in this subfield could be more integrated with each other in the institutions where there is work being done in cognitive science. The same applies to experimental psychology/cognitive science. Organizational psychology, clinical psychology, forensic psychology, and school psychology, for example, are applied fields of cognitive science, as much as engineering is the application of the physical sciences and mathematics in the world, and medicine and veterinary are applications of the biological sciences, chemistry, and other sciences. But they are best kept separated.

A third step is to provide, for the first time, a *Unified Theory of Human Cognition* (UTHC), and its major characteristics, in which the philosophical and theoretical assumptions are plausible and clear. Such UTHC is directed to better characterize the *explanandum* of human cognition in cognitive science. It can be developed by integrated work of the international community of cognitive scientists. This theory can be used as reference for the cognitive standard research program to the extent that it would help to clarify more precisely what exactly human cognition is supposed to refer to, what are the bounds of human cognition and what exactly is the relation between human cognition, environment, neural activity and behavior. The idea behind this standard theory is to give researchers a common and sound basis to better elaborate and refine the theory or to criticize and revise it—similarly to what happens with the standard theories in physics (Laird et al. 2017). UTHC could also guide empirical experimental work and be used for application to practical issues in the field, as well as used for precise comparison in the case an alternative complete theory is formulated. The standard theory should, at the same time, provide criteria as clear and objective as possible for the identification of cognitive entities and systems, cognitive capacities and processes, etc., i.e. it needs to tell us what human cognition is, and what it is not. Moreover, a theory such as this one could provide the international community with the same base to understand and revise central concepts such as 'cognitive computation', 'cognitive information processing' and 'cognitive representation'. Simultaneously, theoretical and philosophical work on these foundations of the theory can be extremely significant for clarification, improvement and revision.

Finally, a fourth step is to provide, also for the first time, a *Unified Theory of Scientific Explanations in Cognitive Science* (UTSECS). This theory can provide a better characterization of the *explanans* in cognitive science, i.e. how successful explanations are constructed in the field and how they ought to be constructed. With such a standard theory, systematic descriptions of successful explanations carried in the broad field can be constructed and normative scientific guidelines based on them can be provided. In this way, what exactly it means to give scientific explanations in the field of cognitive science can be better clarified. The kinds of major explanations in the field can be counted and systematically related. The role played by different fields of inquire on cognition and human cognition and the explanations offered by them can be examined. The standard theory can be compared with other theories of scientific explanations in other successful sciences and be revised and improved accordingly. Thus, UTHC and UTSECS are strictly related: while the first provides information about, roughly, what human cognition is, the second provides information about, roughly, how it can be successfully explained.[10]

6.2 Final Remarks

I presented in this chapter two major theories in the broad field of cognitive science that attempt to provide characterizations and explanations of human cognitive capacities. To this extent they are competitors to MTHC. These two frameworks offer important material for the discussion of central topics concerning the nature of explanations in cognitive science and the nature of human cognitive capacities. More importantly, they also offer significant contributions for the discussion of where MTHC stands as regards such central topics.

Concerning the central issue of scientific explanations in cognitive science, I discussed whether they are constructed in terms of laws or mechanisms, as the advocates of MTHC defend. I discussed important arguments for and against the traditional view that the aim of scientific explanation is to provide natural general laws. Some important arguments concerning this debate are provided by neo-mechanists and their opponents. These arguments were systematically discussed. The final conclusion reached concerning this point is that it is possible to make compatible the idea that scientific explanation in cognitive science is a matter of providing an account of the functions of mechanisms and the classical idea that laws play an important role in explanations in cognitive science. To this extent, I presented an alternative possible and plausible view for understanding the nature of such explanations.

Moreover, I discussed the important issue of the autonomy of psychological explanations. I noted that two of the main claims of advocates of MTHC are (1) that all psychological functional explanations are a kind of mechanistic explanations and (2) the normative claim that all successful mechanistic explanations ought to show how abstract functions are related to neural activity. I provided some examples in order to argue that psychological functional explanation can be strongly autonomous and still successful even in cases where they do not directly relate abstract functions to neural activity. I offered an argument based on the irrelevance of neural information to some psychological explanations of complex human cognitive capacities, such as human conscious complex informal reasoning and decision making. The most relevant information for a psychological explanation in these cases concerns the role that environmental, cultural, social and internal belief systems play in the causation

[10]Of course, a more detailed presentation and elaboration of the most central points of this section would be in order. However, it exceeds the scope of the present book. This can be done in future work.

of a complex patter of human behavior. The advantage of this argument is that it does not depend on highly controversial and unsettled issues about multiple realizability, issues about derivations of laws or theories, and issues concerning the emergence of autonomous causal powers in different levels of neural organization.

As regards the central issue of the nature of human cognitive phenomena, I showed that the diversity and fragmentation in the field regarding major topics is extreme. MTHC is not able to deal with such diversity to the extent that it focusses on a limited number of examples, many of which lie arguably even outside of the field of cognitive science. Moreover, on the important issues of multiple realizability, type neuro-cognitive identity relations, cognitive representations and cognitive computations there is substantial divergence even among proponents of MTHC themselves. Therefore, the unification of the field of cognitive science intended by them cannot be achieved. Many of their views on these issues are highly controversial, misleading or need further elaboration.

Finally, I offered some remarks and suggestions concerning how a more systematic and strong unification of cognitive science can be achieved in the future. In order to establish major guidelines for the scientific work, the field itself needs to be taken seriously. Major examples of scientific successful work being done in all of its major areas of research need to be carefully considered. Besides, more systematic theoretical work needs to be done; for instance, by systematically comparing macro-theories in the field that are in competition to provide the most plausible theory of human cognition. This comparison needs to take into consideration empirical scientific evidence, but also the philosophical foundations of the theories. It could be also helpful to make some improvements and clarifications concerning the relationship between theoretical, experimental and applied cognitive science, in order to make the scientific field more organized and the relationship between the theoretical, experimental and applied dimensions of the science more fruitful. Two other important steps in this direction are the elaboration of a *Unified Theory of Human Cognition* (UTHC) and a *Unified Theory of Scientific Explanations in Cognitive Science* (UTSECS), so that the international community can have a common basis to work and make progress considering well shared unified theories about these crucial matters, criticizing, reviewing and improving them.

References

Adams, F., Aizawa, K.: The Bounds of Cognition. Wiley-Blackwell, Malden, MA (2010)

Aizawa, K.: The biochemistry of memory consolidation: A model system for the philosophy of mind. Synthese **155**, 65–98 (2007)

Aizawa, K.: Neuroscience and multiple realization: A reply to Bechtel and Mundale. Synthese **167**(3), 493–510 (2009)

Aizawa, K.: Multiple realization by compensatory differences. European Journal for Philosophy of Science **3**, 69–86 (2013)

Aizawa, K., Gillett, C.: Levels, individual variation, and massive multiple realization in neurobiology. In: Bickle, J. (ed.) The Oxford Handbook of Philosophy and Neuroscience, pp. 539–581. Oxford University Press, New York (2009)

Aizawa, K., Gillett, C.: The autonomy of psychology in the age of neuroscience. In: Illari, P.M., Russo, F., Williamson, J. (eds.) Causality in the Sciences, pp. 202–223. Oxford University Press, New York (2011)

Anderson, J.R.: Cognitive Psychology and its Implications, 8th edn. Worth Publishe, New York (2015)

Ayer, J.A.: What is a law of nature? Revue Internationale de Philosophie **10**(36), 144–165 (1956)

Bechtel, W.: Levels of description and explanation in cognitive science. Mind. Mach. **4**, 1–25 (1994)

Bechtel, W.: Mental Mechanisms: Philosophical Perspectives on Cognitive Neurosciences. Routledge, New York (2008)

Bechtel, W.: Identity, reduction, and conserved mechanisms: Perspectives from circadian rhythm research. In: Gozzano, S., Hill, C.S. (eds.) New Perspectives on Type Identity: The Mental and the Physical, pp. 43–65. Cambridge University Press, Cambridge (2012)

Bechtel, W.: Explicating top-down causation using networks and dynamics. Philos. Sci. **84**, 253–274 (2017)

Bechtel, W., Mundale, J.: Multiple realizability revisited. Philos. Sci. **66**, 175–207 (1999)

Bechtel, W.: Constructing a philosophy of science of cognitive science. Top. Cogn. Sci. **1**(3), 548–569 (2009a)

Bechtel, W.: Molecules, systems, and behavior: Another view of memory consolidation. In Bickle, J. (ed.), The Oxford Handbook of Philosophy and Neuroscience, pp. 13–40. New York: Oxford University Press (2009b)

Bechtel, W., McCauley, R.N.: Heuristic identity theory (or back to the future): The mind-body problem against the background of research strategies in cognitive neuroscience. In: Hahn, M., Stoness, S.C. (eds.) Proceedings of the 21st Annual Meeting of the Cognitive Science Society (pp. 67–72). Mahwah, NJ: Lawrence Erlbaum Associates (1999)

Bechtel, W., Shagrir, O.: The non-redundant contributions of Marr's three levels of analysis for explaining information-processing mechanisms. Topics in Cognitive Science **7**, 312–322 (2015)

Bechtel, W., Wright, C.D.: What is psychological explanation? In: Symons, J., Calvo, P. (eds.) The Routledge Companion to Philosophy of Psychology, pp. 113–130. Routledge and Taylor & Francis Group, New York (2009)

Beck, A. T. & Emery, G. (with Greenberg, R. L.): Anxiety Disorders and Phobias: A Cognitive Perspective. New York: Basic Books (1985)

Bermúdez, J.L.: Philosophy of Psychology: A Contemporary Introduction. Routledge, New York (2005)

Bermúdez, J.L.: Cognitive Science: An Introduction to the Science of the Mind, 2nd edn. Cambridge University Press, Cambridge (2014)

Bickle, J.: Philosophy and Neuroscience: A Ruthlessly Reductive Account. Kluwer Academic Press, Norwell, MA (2003)

Bickle, J.: Reducing mind to molecular pathways: Explicating the reductionism implicit in current cellular and molecular neuroscience. Synthese **151**, 411–434 (2006)

Bickle, J.: A brief history of neuroscience's actual influences on mind–brain reductionism. In: Gozzano, S., Hill, C.S. (eds.) New Perspectives on Type Identity: The Mental and the Physical, pp. 88–110. Cambridge University Press, Cambridge (2012)

Bickle, J.: Multiple realizability. The Stanford Encyclopedia of Philosophy (Spring 2016 Edition). In: Zalta, E.N. (ed.) (2020). https://plato.stanford.edu/archives/spr2016/entries/multiple-realizability/

Block, N.: On a confusion about the function of consciousness. Behav. Brain Sci. **18**, 227–247 (1995)

Block, N.: Anti-reductionism slaps back. Philos. Perspect. **11**, 107–132 (1997)

Boone, W., Piccinini, G.: Cognitive neuroscience revolution. Synthese (2015). https://doi.org/10.1007/s11229-015-0783-4

Braddon-Mitchell, D., Jackson, F.: Philosophy of Mind and Cognition, 2nd edn. Blackwell Publishing, Malden, MA (2007)
Cat, J.: The unity of Science. The Stanford Encyclopedia of Philosophy (Fall 2017 Edition). In: Zalta, E.N. (ed.) (2017). https://plato.stanford.edu/archives/fall2017/entries/scientific-unity/
Chipman, S. (ed.): The Oxford Handbook of Cognitive Science. Oxford University Press, Oxford (2017)
Clark, A.: Mindware: An Introduction to the Philosophy of Cognitive Science, 2nd edn. Oxford University Press, Oxford (2014)
Clark, D.A., Beck, A.T.: Cognitive Therapy of Anxiety Disorders: Science and Practice. Guilford Press, New York (2010)
Craver, C.F.: Explaining the Brain: Mechanisms and the Mosaic Unity of Neuroscience. Oxford University Press, Oxford (2007)
Craver, C.F., Tabery, J.: Mechanisms in Science. The Stanford Encyclopedia of Philosophy (Spring 2017 Edition). In Zalta, E.N. (ed.) (2015). https://plato.stanford.edu/archives/spr2017/entries/science-mechanisms/
Craver, C.F., Kaiser, M.I.: Mechanisms and laws: Clarifying the debate. In: Chao, H., Chen, S., Millstein, R.L. (eds.) Mechanism and Causality in Biology and Economics, pp. 125–146. Springer, Dordrecht (2013)
Cummins, R.: The Nature of Psychological Explanation. MIT Press, Cambridge, MA (1983)
Cummins, R.: "How does it work" versus "what are the laws?": Two conceptions of psychological explanation. In: Keil, F., Wilson, R.A. (eds.) Explanation and Cognition, pp. 117–145. MIT Press, Cambridge, MA (2000)
Eronen, M.: Hypothetical identities: Explanatory problems for the explanatory argument. Philos. Psychol. **27**(4), 571–582 (2014)
Evans, T.M., et al.: Evidence for a mental health crisis in graduate education. Nat. Biotechnol. **36**(3), 283–284 (2018)
Eysenck, M.W., Keane, M.T.: Cognitive Psychology: A Student's Handbook, 7th edn. Psychology Press, New York (2015)
Fechner, G.T.: Elemente der Psychophysik, Erster Theil. Zweite Auflage. Breitkopf & Hartel, Leipzig (1889)
Figdor, C.: Neuroscience and the multiple realization of cognitive functions. Philos. Sci. **77**(3), 419–456 (2010)
Fodor, J.: Psychological Explanation: An Introduction to the Philosophy of Psychology. Random House, New York (1968)
Fodor, J.: Special sciences. Synthese **28**(2), 97–115 (1974)
Fodor, J.: A Theory of Content and Other Essays. The MIT Press, Cambridge, MA (1990)
Fodor, J.: You can fool some of the people all of the time, everything else being equal; Hedged laws and psychological explanation. Mind **100**, 19–34 (1991)
Fodor, J.: Special sciences: Still autonomous after all these years. Philos. Perspect. **11**, 149–163 (1997)
Fodor, J., Pylyshyn, Z.: Connectionism and cognitive architecture: A critical analysis. Cognition **28**, 3–71 (1988)
Gardner, H.: The Mind's New Science: A History of the Cognitive Revolution. Basic Books, New York (1985)
Garson, J.: Connectionism. The Stanford Encyclopedia of Philosophy (Winter 2016 Edition). In: Zalta, E.N. (ed.) (2019). https://plato.stanford.edu/archives/win2016/entries/connectionism/
Goldstein, B.: Cognitive Psychology: Connecting Mind, Research and Everyday Experience, 4th edn. Cengage Learning, Stamford, CT (2015)
Grush, R., Damm, L.: Cognition and the Brain. In: Margolis, E., Samuels, R., Stich, S. (eds.) The Oxford Handbook of Philosophy of Cognitive Science, pp. 273–290. Oxford University Press, New York (2012)
Harnish, R.M.: Minds, Brains, Computers: An Historical Introduction to the Foundations of Cognitive Science. Blackwell Publishers, Malden (2002)

Hempel, C.G., Oppenheim, P.: Studies in the logic of explanation. In: Hempel, C.G. (eds.) Aspects of Scientific Explanation and Other Essays in the Philosophy of Science (pp. 245–290). New York: Free Press (Originally published in 1948) (1965)

Hofmann, S.G.: An Introduction to Modern CBT. Wiley-Blackwell, Malden, MA (2012)

Jackson, F.: What Mary didn't know. J. Philos. **83**, 291–295 (1986)

Jones, P.J., Park, S.Y., Lefevor, G.T.: Contemporary college student anxiety: The role of academic distress, financial stress, and support. J. Coll. Couns. **21**, 252–264 (2018). https://doi.org/10.1002/jocc.12107

Kalat, J.W.: Introduction to Psychology, 9th edn. Wadsworth Cengage Learning, Belmont, CA (2011)

Kim, J.: Multiple realization and the metaphysics of reduction. Philos. Phenomenological Res. **52**, 1–26 (1992)

Kim, J.: Physicalism, or Something Near Enough. Princeton University Press, Princeton (2005)

Kim, J.: Philosophy of Mind, 3rd edn. Westview Press, Boulder (2011)

Koons, R.C., Bealer, G. (eds.): The Waning of Materialism. Oxford University Press, Oxford (2010)

Korsgaard, C.: The Sources of Normativity. Cambridge University Press, Cambridge (1996)

Laird, J.E., Lebiere, C., Rosenbloom, P.S.: A standard model for the mind: Toward a common computational framework across artificial intelligence, cognitive science, neuroscience, and robotics. AI Mag. **38**(4), 13–26 (2017). https://doi.org/10.1609/aimag.v38i4.2744

Leuridan, B.: Can mechanisms really replace laws of nature? Philos. Sci. **77**, 317–340 (2010)

Luria, A.R.: Cognitive Development: Its Cultural and Social Foundations. Harvard University Press, Cambridge, MA (Originally published in Russian, 1974) (1976)

Margolis, E., Samuels, R., Stich, S. (eds.): The Oxford Handbook of Philosophy of Cognitive Science. Oxford University Press, New York (2012)

Marr, D.: Vision: A computational investigation into the human representation and processing of visual information. In: Freeman, W.H. (ed.) San Francisco (1982)

McLaughlin, B.: Consciousness, type physicalism, and inference to the best explanation. Philos. Issues **20**, 266–304 (2010)

Mele, A.: Unconscious decisions and free will. Philos. Psychol. **26**(6), 777–789 (2013)

Mitchell, S.D.: Pragmatic laws. Philos. Sci. **64**, S468–S479 (1997)

Nagel, T.: What is it like to be a bat? Philos. Rev. **83**, 435–456 (1974)

Ochsner, K.N., Kosslyn, S.M. (eds.): The Oxford Handbook of Cognitive Neuroscience, vol. 1. Oxford University Press, Oxford (2014)

Papineau, D.: Thinking About Consciousness. Oxford University Press, Oxford (2002)

Pedrelli, P., et al.: College Students: Mental Health Problems and Treatment Considerations. Acad. Psychiatry (2014). https://doi.org/10.1007/s40596-014-0205-9

Piccinini, G.: Computationalism. In: Margolis, E., Samuels, R., Stich, S. (eds.) The Oxford Handbook of Philosophy of Cognitive Science, pp. 222–249. Oxford University Press, New York (2012)

Piccinini, G., Craver, C.: Integrating psychology and neuroscience: Functional analyses as mechanism sketches. Synthese **183**(3), 283–311 (2011)

Piccinini, G., Maley, C.: The metaphysics of mind and the multiple sources of multiple realizability. In: Sprevak, M., Kallestrup, J. (eds.) New Waves in the Philosophy of Mind, pp. 125–152. Palgrave Macmillan, New York (2014)

Pietroski, P., Rey, G.: When other things aren't equal: Saving *ceteris paribus* laws from vacuity. Br. J. Philos. Sci. **46**, 81–110 (1995)

Pitt, D.: Mental representation. The Stanford Encyclopedia of Philosophy (Spring 2017 Edition). In: Zalta, E.N. (ed.) (2020). https://plato.stanford.edu/archives/spr2017/entries/mental-representation/

Polger, T.: Evaluating the evidence for multiple realization. Synthese **167**, 457–472 (2009)

Polger, T.W., Shapiro, L.A.: The Multiple Realization Book. Oxford University Press, Oxford (2016)

Putnam, H.: Mind Language and Reality: Philosophical Papers, vol. 2. Cambridge University Press, Cambridge (1975)

Putnam, H.: The nature of mental states. In: Putnam, H. (ed.) Mind Language and Reality: Philosophical Papers, vol. 2, pp. 429–440. Cambridge University Press, Cambridge (Originally published as "Psychological Predicates" in 1967) (1975)
Quek, T.T. et al.: The global prevalence of anxiety among medical students: A meta-analysis. Int. J. Environ. Res. Public Health **16**. https://doi.org/10.3390/ijerph16152735 (2019)
Reisberg, D. (ed.): The Oxford Handbook of Cognitive Psychology. Oxford University Press, Oxford (2013)
Rescorla, M.: The computational theory of mind. The Stanford Encyclopedia of Philosophy (Spring 2017 Edition). In: Zalta, E.N. (ed.) (2020). https://plato.stanford.edu/archives/spr2017/entries/computational-mind/
Reutlinger, A., Schurz, G., Hüttemann, A.: Ceteris Paribus Laws. The Stanford Encyclopedia of Philosophy (Spring 2017 Edition). In Zalta, E.N. (ed.) (2015). https://plato.stanford.edu/archives/spr2017/entries/ceteris-paribus/
Roeckelein, J.E.: Citations of 'laws' and 'theories' in textbooks across 112 years of psychology. Psychol. Rep. **77**, 163–174 (1996)
Roedinger, H.L.: Relativity of remembering: Why the laws of memory vanished. Annu. Rev. Psychol. **59**, 225–254 (2008)
Rosenberg, A.: Making mechanism interesting. Synthese (2015). https://doi.org/10.1007/s11229-015-0713-5
Shapiro, L.: Multiple realizations. J. Philos. **97**, 635–654 (2000)
Skinner, B.F.: Why I am not a cognitive psychologist. Behaviorism **5**(2), 1–10 (1977)
Skinner, B.F.: Can psychology be a science of mind? Am. Psychol. **45**, 1206–1210 (1990)
Smart, J.: Sensations and brain processes. Philos. Rev. **68**(2), 141–156 (1959)
Sober, E.: Two outbreaks of lawlessness in recent philosophy of biology. Philos. Sci. **64**, S458–S467 (1997)
Sternberg, R.J., Sternberg, K.: Cognitive Psychology, 6th edn. Wadsworth, Cengage Learning, Belmont, CA (2012)
Stoljar, D.: Physicalism. Routledge, New York (2010)
Stoljar, D.: Physicalism. The Stanford Encyclopedia of Philosophy (Spring 2015 Edition). In: Zalta, E.N. (ed.) (2015). http://plato.stanford.edu/archives/spr2015/entries/physicalism/
Teigen, K.H.: One hundred years of laws in psychology. Am. J. Psychol. **115**(1), 103–118 (2002)
Thagard, P.: Mind: Introduction to Cognitive Science, 2nd edn. MIT Press, Cambridge, MA (2005)
Thagard, P.: Hot Thought: Mechanisms and Applications of Emotional Cognition. MIT Press, Cambridge, MA (2006)
Thagard, P.: Cognitive Science. The Stanford Encyclopedia of Philosophy (Winter 2018 Edition). In Zalta, E.N. (ed.), forthcoming (2018). https://plato.stanford.edu/archives/win2018/entries/cognitive-science/
Van Gulick, R.: Nonreductive materialism and the nature of intertheoretical constraint. In: Beckermann, A., Flohr, H., Kim, J. (eds.) Emergence or Reduction? Essays on the Prospects of Nonreductive Physicalism, pp. 157–179. Walter de Gruyter, Berlin (1992)
Van Gulick, R.: Consciousness. The Stanford Encyclopedia of Philosophy (Spring 2014 Edition). In: Zalta, E.N. (ed.) (2014). http://plato.stanford.edu/archives/spr2014/entries/consciousness/
Von Eckardt, B., Poland, J.S.: Mechanism and explanation in cognitive neuroscience. Philos. Sci. **71**(5), 972–984 (2004)
Ward, J.: The Student's Guide to Cognitive Neuroscience, 3rd edn. Psychology Press, New York (2015)
Weiskopf, D.: The functional unity of special science kinds. Br. J. Philos. Sci. **62**, 233–258 (2011)
Weiskopf, D., Adams, F.: An Introduction to the Philosophy of Psychology. Cambridge University Press, Cambridge (2015)
Woodward, J.: There is no such thing as a *ceteris paribus* law. Erkenntnis **57**, 303–328 (2002)
Woodward, J.: Scientific explanation. The Stanford Encyclopedia of Philosophy (Fall 2017 Edition). In: Zalta, E.N. (ed.) (2014). https://plato.stanford.edu/archives/fall2017/entries/scientific-explanation/

Zednik, C.: Mechanisms in cognitive science. In: Glennan, S., Illari, P. (eds.) The Routledge Handbook of Mechanisms and Mechanical Philosophy, pp. 389–400. Routledge, London (2018)

Conclusion

MTHC has been in many ways ground-breaking for the sciences investigating human cognition. It provides, for the first time, a clearer and more systematic analysis of the concept of 'mechanism' which is overwhelmingly used in contemporary cognitive science and neuroscience. Moreover, the account clarifies more exactly what it is to construct mechanistic scientific explanations in many fields of science and to what extent this kind of scientific explanation is present in contemporary cognitive science and neuroscience. The theory also proposes an explanatory pluralist (multi-level) integration of neuro-cognitive phenomena, on the one hand, and cognitive science and neuroscience, on the other, promoting important debates concerning the unity and integration of cognitive science and its further integration with relevant portions of neuroscience. In this sense, the theory attempts to provide clearer philosophical and theoretical foundations for the field of cognitive science, but especially for the field of cognitive neuroscience. This philosophical and theoretical articulation of the foundations in these sciences permits a more reasonable and fair judgment about the merits of different research programs present in different scientific fields currently interested in human cognition. And since the neo-mechanistic framework is very influential, it also provides important guidelines for a vast amount of work being currently done in cognitive neuroscience and cognitive science, as well. The theory, therefore, can be arguably considered one of the major current philosophical and scientific accounts providing understanding about human cognition.

However, the detailed and rigorous analysis of the theory carried out in the present work shows that it has major internal inconsistencies and theoretical limitations. The very formulation of central elements of the theory is not entirely clear. I found considerable disagreement among leading neo-mechanists concerning the most plausible formulation of MTHC regarding fundamental issues. There is a great amount of disagreement concerning taxonomy and interpretation of the most central concepts; for instance, the concepts of 'behavior', 'function', 'activities', 'operations', 'cognitive representation', and 'cognitive computation'. Often it is hard to point out whether the differences in details concerning the meaning of these concepts is significant or not. There is also a great amount of controversy concerning, for example, the issue of mechanistic levels of organization and the relevant levels required to explain

D. A. Leite, *The Twenty-First Century Mechanistic Theory of Human Cognition*,
Cognitive Systems Monographs 41, https://doi.org/10.1007/978-3-030-63680-7

human cognition, which are counted differently by different major proponents of MTHC. There is also controversy on the central issue of multiple realizability of human cognitive capacities—some major neo-mechanists argue forcefully against it and others accept it flatly as plausible. Another major controversy concerns the role played by mind-brain type-type identity theory in the framework. Again, some neo-mechanists accept it, while others oppose it. Moreover, there is controversy concerning the equally central issue of epistemological neuro-cognitive reduction—some major authors advocate a particular view on epistemological reduction, while others completely oppose epistemological reductionism in cognitive and neural sciences. The theory needs, therefore, further development and better elaboration, especially regarding these central issues.

Another major problem within the neo-mechanistic theory is related to the issue of pluralistic explanatory neuro-cognitive scientific integration. Leading proponents of MTHC are committed to this integrative position, but this commitment faces serious shortcomings. The strategy of the advocates of the theory is to build the integration starting from a sort of 'middle level'. In this sense, they intend to unify bottom-up lower-level frameworks with higher-level top-down frameworks. On the one hand, they attempt to unify work on neural networks with traditional ideas about cognitive information processing, representation and sometimes also computation. On the other hand, MTHC also attempts to build relations with theoretical proposals concerning relevant portions of neuroscience, such as the strong neuro-cognitive reductionism of Bickle, based on the field of (cognitive) molecular and cellular neuroscience. While the neo-mechanistic framework attempts to construct such integration, it also tries to avoid the strong reductionist implications.

However, my analysis shows that MTHC cannot avoid strong explanatory neuro-cognitive reductionism and it collapses into a strong reductionist framework. MTHC and the cognitive molecular and cellular framework (MCTHC) are concerned with the same phenomena—they simply use different vocabularies. These different vocabularies are not a matter of different explanatory levels. To put it roughly, neural networks are made of cells and cells are made of subcellular components—this can hardly be challenged. MTHC emphasizes more complex neural networks and their properties as a system, while MCTHC emphasizes subcellular components, such as molecular processes, and single cells interactions in terms of synapses and neurotransmitters. Some relations between these levels cannot yet be fully understood because we still need to acquire better knowledge of many functions of the brain at both levels and also because we lack more accurate methodology. However, with the improvements of neuroscientific methodologies, as well as with improvements in the taxonomy of brain regions—making it as systematic, clear, precise and standard as possible –, it appears to be very likely that these methodological and theoretical gaps can be eliminated.

Ironically, this is a coherent way in which the neo-mechanistic framework can provide some substantial scientific unification for the cognitive and neural sciences. MTHC can assimilate strong neuro-cognitive explanatory reductionism and incorporate MCTHC. This integration can provide a comprehensive theory of the putative neural basis of natural human cognition. In this way, the different vocabularies can be related as much as the theoretical and methodological advancements permit. This comprehensive theory of the neural basis of human cognition can be developed in a comprehensive field of cognitive neuroscience that can have two major divisions: one dealing with the molecular and cellular bases of human cognitive processes; and the other dealing with (complex) neural network systems as bases of human cognitive processes.

The analysis carried out in this work also shows that MTHC can incorporate as well the dynamical systems theory of human cognition (DSTHC), to the extent that the complex organization that best characterizes complex systems are part of the possible organization that biological mechanisms can have. Moreover, throughout the years, MTHC made amendments in its strictness concerning decomposability of complex systems due to the criticism of dynamicists. Decomposition becomes merely a heuristic principle, what can make dynamical explanations for complex biological systems part of MTHC without any conflict. However, strong neuro-cognitive reductionism will still be present here, since the unpredictability of complex systems does not pose a problem to the descriptions of the neural complex phenomena at lower-levels.

Based on these results, one can argue that with some central and peripheral modifications, MTHC is able to provide some initial unification of MCTHC and DSTHC. These unifications, together with the central theoretical elements of the main theory, need, evidently, better elaboration in order to be proposed in a well-articulated fashion. This is a task that can be undertaken by the contemporary neo-mechanists interested in cognitive science.

Concerning the theory of situated human cognition, the analysis shows that it presents major problems for the intentions of neo-mechanists concerning integration in cognitive science. The original and relevant version of the situated theory, i.e. TSHC, states a thesis on the nature of human cognition opposed to the one defended by proponents of MTHC. In the point of view of influential neo-mechanists, human cognitive capacities depend only upon neural components to be realized. However, in the view of the advocates of the strong situated view of human cognition, the components realizing some of the human cognitive capacities are not exclusively neural—they can also be bodily or environmental components, natural or artificial. Since the advocates of MTHC are not able to present definitive arguments in order to show either the compatibility between the theories, or that TSHC is incorrect, this latter framework still presents an alternative view on human cognition. This outcome does not show that TSHC is a correct or plausible theory. It does show, nevertheless, that the ambition of MTHC to unify the field of cognitive science (where there is major work being done under the guidance of TSHC) cannot be fulfilled. Since both theories have internal problems and inconsistencies, to the extent that it is yet difficult to accurately determine their degree of plausibility in comparison to each other, it remains the case that they are two alternative views on the nature of human cognitive capacities competing to provide the best explanation in the broad field of cognitive science. Some authors will tend to one side; while others will tend to the other side—while no particular set of objective and plausible criteria in order to decide which is the best comprehensive theory is presented.

The mechanistic theory attempts to build relations with theories of human cognition at a higher-level of abstraction as well. This is the case especially with the classic computational theory of human cognition (CTHC), but it happens also with the classic belief-desire theory of human cognition (BDTHC). The major strategy presented by advocates of MTHC is to use central notions of these frameworks to build its integrative project. The analysis carried out in this work shows, however, that, in this respect, the neo-mechanistic theory also presents significant shortcomings and inconsistencies. The theory requirements on scientific explanations in cognitive science impose unnecessary and distorted restrictions in the way many explanations of conscious complex cognitive capacities and related behaviors are constructed in the field. Moreover, on crucial issues in cognitive science related to the nature of cognitive phenomena there is substantial disagreement even among the most influential proponents of MTHC. The advocates of the theory diverge substantially concerning multiple realizability of cognitive capacities, type neuro-cognitive identity relations, cognitive representations, and cognitive computation. Thus, it is also difficult to make an assessment of the negative and positive aspects of the theory, since it can have very different formulations. In addition, the neo-mechanists present views on these issues that are extremely controversial, or that are obscure and need further elaboration. Accordingly, the project of unification intended by influential neo-mechanists in cognitive science is highly compromised by all the substantial divergence found among neo-mechanists themselves, as well as by the shortcomings of some of their positions on central issues.

From these results, it is possible to drawn the following further conclusions. MTHC, thus, cannot provide the unification promised in a coherent and consistent form. Given this outcome, one can consider at least four major scientific theories of human cognition competing in the contemporary broad field of cognitive science. These theories are: MTHC (after integration with MCTHC and DSTHC), TSHC, CTHC, and BDTHC. MTHC does, therefore, a reasonable theoretical work unifying a great portion of theories developed in fields of neuroscience interested in explaining human cognition—if it is formulated with the incorporation of strong neuro-cognitive explanatory reduction and dynamical explanations of complex biological systems. However, especially when it comes to more complex human cognitive capacities, such as conscious complex informal reasoning and conscious complex decision making, there are particular complex issues that poses major limitations to the framework.

Therefore, the central conclusion of this work is that MTHC promises more than what it is actually able to provide. It has enormous ambitions, but a very limited capacity to achieve them. It aims at an integration within cognitive science and, even more, at a further integration between cognitive science and relevant portions of neuroscience, but it does not offer enough concrete examples or a single clear and well-articulated proposal of how that could be achieved in all the dimensions required by such ambitious integrative project. The problems for an integration of this kind are still many, as the present work shows, and some of the major problems are not overcome by MTHC. What the theory is able to offer so far are general ideas or the indication of possibilities concerning such integration. Ultimately, thus, the claim that the mechanistic theory of human cognition provides a ‘new revolution’ in the cognitive and neural sciences is inconsistent and untenable.

This final result, together with the discussions carried out in this work about central issues concerning human cognition and cognitive science, suggest that it will probably take still many years until we are able to find more solid and systematic theoretical integration in the field of cognitive science. However, this analysis also promotes the recognition that we are dealing with an extremely difficult and complex problem and despite the fact that we already know a great amount about it, there is still a great amount that needs to be known.

This recognition may lead us to a genuine pluralism in cognitive science, in which the different macro-theories compete fairly to provide the best explanation. In order to understand this enormously complex topic of human cognition, it is important to consider the plurality of plausible views we all have so far been able to present. The theoretical different points of view in various issues are too many. The central concepts need to be better clarified and the different points of view compared systematically. In this way we can develop a systematic form of organizing and integrating cognitive science and the macro-theories present in the field currently. And although it is not possible yet to point out clearly which theory in cognitive science is the most plausible, one can systematically analyze the theories in offer and compare their positive and negative aspects, coming up ideally with a new synthesis that unifies all their advantages and leave the disadvantages. This scientific and philosophical strategy can effectively promote the scientific progress in cognitive science and improve our understanding of human cognition.

The manufacturer's authorised representative in the EU is Springer Nature Customer Service Centre GmbH, Europaplatz 3, 69115 Heidelberg, Germany. If you have any concerns regarding our products, please contact ProductSafety@springernature.com

Printed and bound by CPI Group (UK) Ltd, Croydon, CR0 4YY
30/09/2026
02225696-0009